my **revision** notes

AQA GCSE CHEMISTRY

For A* to C

Philip Dobson

Philip Allan Updates, an imprint of Hodder Education, an Hachette UK company, Market Place, Deddington, Oxfordshire OX15 0SE

Orders

Bookpoint Ltd, 130 Milton Park, Abingdon, Oxfordshire OX14 4SB

tel: 01235 827827

fax: 01235 400401

e-mail: education@bookpoint.co.uk

Lines are open 9.00 a.m.–5.00 p.m., Monday to Saturday, with a 24-hour message answering service. You can also order through Philip Allan Updates website: www.philipallan.co.uk

ISBN 978-1-4441-2082-0

Impression number 5 4 3 2

Year 2016 2015 2014 2013 2012

Cover photo reproduced by permission of Inside Vision Lab/Fotolia
Photos on page 80 courtesy of John Olive

Printed in India

Get the most from this book

This book will help you revise units Chemistry 1–3 of the new AQA specification. You can use the contents list on pages 2 and 3 to plan your revision, topic by topic. Tick each box when you have:

1. revised and understood a topic
2. tested yourself
3. checked your answers and practised exam questions online

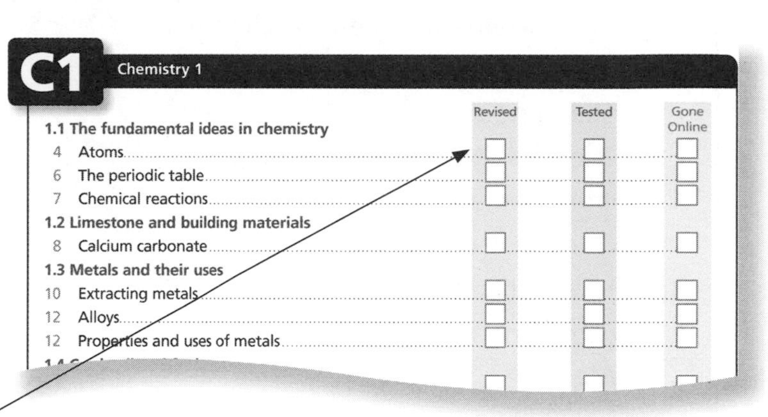

C1 Chemistry 1

	Revised	Tested	Gone Online
1.1 The fundamental ideas in chemistry			
4 Atoms	☐	☐	☐
6 The periodic table	☐	☐	☐
7 Chemical reactions	☐	☐	☐
1.2 Limestone and building materials			
8 Calcium carbonate	☐	☐	☐
1.3 Metals and their uses			
10 Extracting metals	☐	☐	☐
12 Alloys	☐	☐	☐
12 Properties and uses of metals	☐	☐	☐

You can also keep track of your revision by ticking off each topic heading through the book. You may find it helpful to add your own notes as you work through each topic.

Tick to track your progress

examiner tips

Throughout the book there are exam tips that explain how you can boost your final grade.

Higher tier

Some parts of the AQA specification are tested only on higher-tier exam papers. These sections are highlighted using a solid yellow strip down the side of the page.

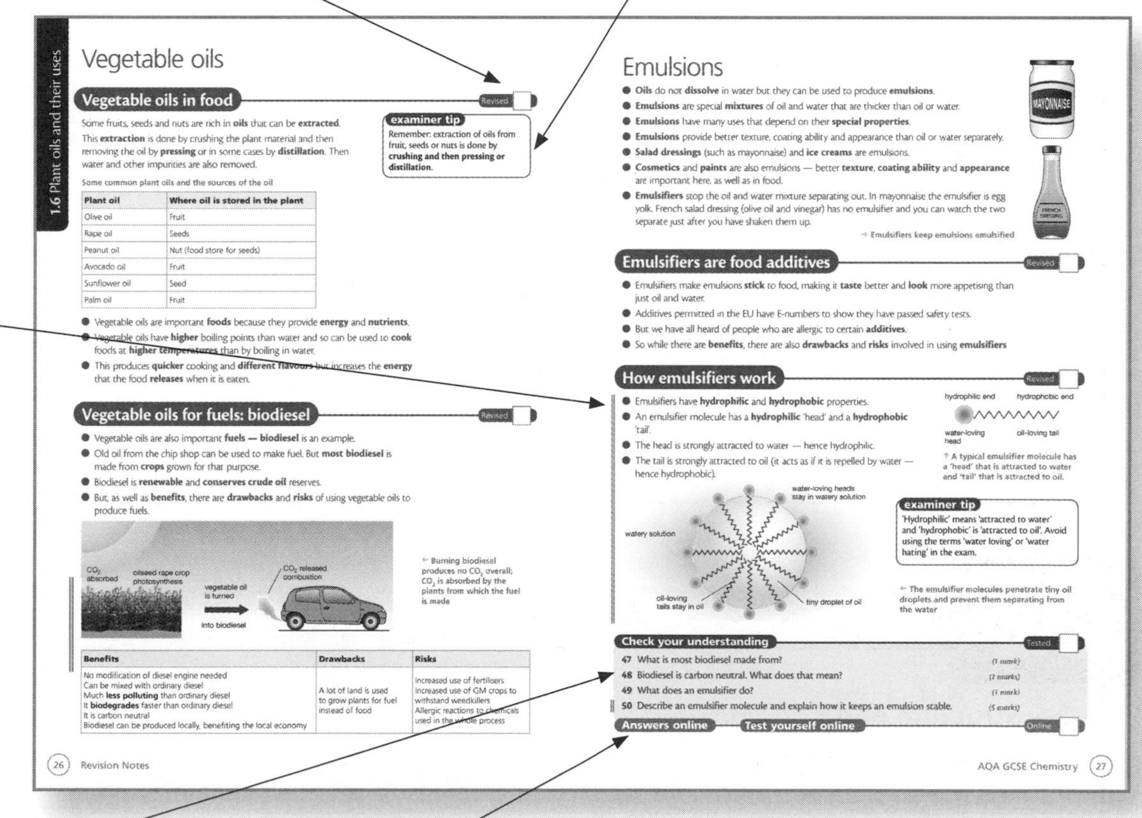

1.6 Plant oils and their uses

Vegetable oils

Vegetable oils in food Revised

Some fruits, seeds and nuts are rich in **oils** that can be **extracted**. This **extraction** is done by crushing the plant material and then removing the oil by **pressing** or in some cases by **distillation**. Then water and other impurities are also removed.

examiner tip
Remember: extraction of oils from fruit, seeds or nuts is done by **crushing and then pressing or distillation.**

Some common plant oils and the sources of the oil

Plant oil	Where oil is stored in the plant
Olive oil	Fruit
Rape oil	Seeds
Peanut oil	Nut (food store for seeds)
Avocado oil	Fruit
Sunflower oil	Seed
Palm oil	Fruit

- Vegetable oils are important **foods** because they provide **energy** and **nutrients**.
- Vegetable oils have **higher** boiling points than water and so can be used to **cook** foods at **higher temperatures** than by boiling in water.
- This produces **quicker** cooking and **different flavours** but increases the **energy** that the food **releases** when it is eaten.

Vegetable oils for fuels: biodiesel Revised

- Vegetable oils are also important **fuels — biodiesel** is an example.
- Old oil from the chip shop can be used to make fuel. But **most biodiesel** is made from **crops** grown for that purpose.
- Biodiesel is **renewable** and **conserves crude oil** reserves.
- But, as well as **benefits**, there are **drawbacks** and **risks** of using vegetable oils to produce fuels.

Burning biodiesel produces no CO_2 overall; CO_2 is absorbed by the plants from which the fuel is made

Benefits	Drawbacks	Risks
No modification of diesel engine needed Can be mixed with ordinary diesel Much **less polluting** than ordinary diesel It **biodegrades** faster than ordinary diesel It is carbon neutral Biodiesel can be produced locally, benefiting the local economy	A lot of land is used to grow plants for fuel instead of food	Increased use of fertilisers Increased use of GM crops to withstand weedkillers Allergic reactions to chemicals used in the whole process

26 Revision Notes

Emulsions

- **Oils** do not **dissolve** in water but they can be used to produce **emulsions**.
- **Emulsions** are special **mixtures** of oil and water that are thicker than oil or water.
- **Emulsions** have many uses that depend on their **special properties**.
- **Emulsions** provide better texture, coating ability and appearance than oil or water separately.
- **Salad dressings** (such as mayonnaise) and **ice creams** are emulsions.
- **Cosmetics** and **paints** are also emulsions — better **texture**, **coating ability** and **appearance** are important here, as well as in food.
- **Emulsifiers** stop the oil and water mixture separating out. In mayonnaise the emulsifier is egg yolk. French salad dressing (olive oil and vinegar) has no emulsifier and you can watch the two separate just after you have shaken them up.

Emulsifiers keep emulsions emulsified

Emulsifiers are food additives Revised

- Emulsifiers make emulsions **stick** to food, making it **taste** better and **look** more appetising than just oil and water.
- Additives permitted in the EU have E-numbers to show they have passed safety tests.
- But we have all heard of people who are allergic to certain **additives**.
- So while there are **benefits**, there are also **drawbacks** and **risks** involved in using **emulsifiers**

How emulsifiers work Revised

- Emulsifiers have **hydrophilic** and **hydrophobic** properties.
- An emulsifier molecule has a **hydrophilic** 'head' and a **hydrophobic** 'tail'.
- The head is strongly attracted to water — hence hydrophilic.
- The tail is strongly attracted to oil (it acts as if it is repelled by water — hence hydrophobic).

A typical emulsifier molecule has a 'head' that is attracted to water and 'tail' that is attracted to oil.

examiner tip
'Hydrophilic' means 'attracted to water' and 'hydrophobic' is 'attracted to oil'. Avoid using the terms 'water loving' or 'water hating' in the exam.

The emulsifier molecules penetrate tiny oil droplets and prevent them separating from the water

Check your understanding Tested

47 What is most biodiesel made from? *(1 mark)*
48 Biodiesel is carbon neutral. What does that mean? *(2 marks)*
49 What does an emulsifier do? *(1 mark)*
50 Describe an emulsifier molecule and explain how it keeps an emulsion stable. *(5 marks)*

Answers online **Test yourself online** Online

AQA GCSE Chemistry 27

Check your understanding

Use these questions at the end of each section to make sure that you have understood every topic.

Go online

Go online to check your answers at **www.therevisionbutton.co.uk/myrevisionnotes**.

Here you can also find extra exam questions for topics as well as podcasts to support you when getting ready for the big day.

Contents and revision planner

C1 Chemistry 1

C2 Chemistry 2

C3 Chemistry 3

Atoms

- Atoms and elements are the building blocks of chemistry.
- Atoms contain protons, neutrons and electrons.
- When elements react they produce compounds.

Atomic structure

Revised

- All substances are made of atoms.
- An **element** is a substance made of only one sort of atom.

examiner tip

Learn this definition of an element — for full marks you need to mention 'element' in the definition.

He
TRANSITION METALS
NON-METALS
Ti Cr Mn Fe Ni Cu Zn
REACTIVE METALS
Ag
POOR METALS
Pt Au

examiner tip

Know where the metals and non-metals appear in the periodic table.

← Non-metals are on the right-hand side of the periodic table

- There are about 100 different **elements**.
- **Elements** are shown in the **periodic table**.
- The **groups** (columns) contain **elements** with similar properties.
- A **chemical symbol** represents one atom of an element, for example O represents an atom of oxygen; Na represents an atom of sodium.
- Atoms have a small central **nucleus** made up of **protons** and **neutrons** around which there are **electrons**.

electron
nucleus

↑ An atom: a nucleus surrounded by electrons

The table summarises the charges and location of sub-atomic particles.

The charges and positions within atoms of protons, neutrons and electrons

Particle	Relative charge	Position within atom
Proton	+1	Nucleus
Neutron	0	Nucleus
Electron	−1	In space outside the nucleus

- In an **atom**, the number of **electrons** equals the number of **protons**. Atoms have no electrical charge.
- Atoms of a particular element have the **same number of protons**.
- Atoms of **different** elements have **different** numbers of **protons**.
- The number of **protons** in an atom is its **atomic number**.
- The **sum** of protons and neutrons in an atom is its **mass number**.

examiner tip

Make sure you know the difference between atomic number and mass number.

examiner tip

Energy levels or **shells**? The examiner does not mind!

Electronic structures of atoms

Revised

Electrons occupy particular **energy levels**, sometimes called **shells**.

Each electron in an atom is in a particular **energy level**.

The electrons in an atom occupy the **lowest available** energy levels (the innermost **shells**).

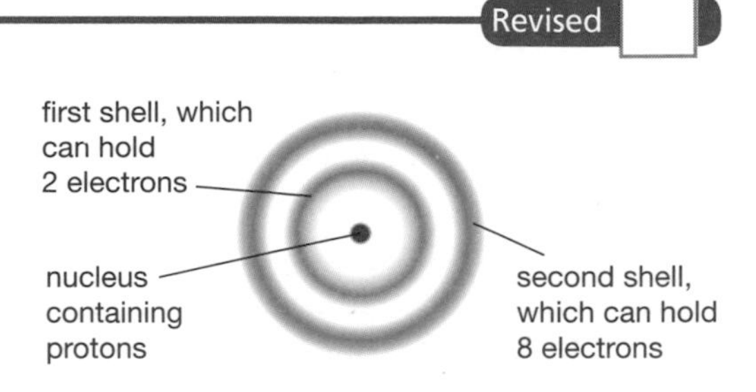

→ A model for the arrangement of electrons in the first and second shells

Filling electron shells — innermost shells first

Revised

Work through this with a periodic table to hand:

1 Starting with an atom of hydrogen, electrons fill the shell closest to the nucleus first.

2 When that shell is full (two electrons, helium) the next electron goes into the next shell out.

3 This second shell can hold up to eight electrons, enough for the eight elements across the first row of the periodic table.

4 For sodium (atomic number 11), the 11th electron has to go into the third shell (energy level).

5 This shell can also hold up to eight electrons and so we again move across the period to argon until the third shell is full.

6 Electrons then start to fill the fourth shell in potassium and calcium.

examiner tip

You will have a periodic table in the exam — use it to work out electronic structures.

The **electronic structure** of the first 20 elements of the periodic table can be represented as diagrams, as shown here for helium, sodium and calcium, or in a simple form such as Na (2,8,1).

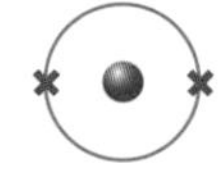

an atom of helium, He (2)

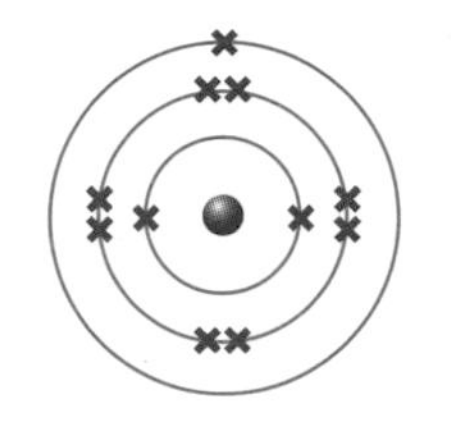

an atom of sodium, Na (2,8,1)

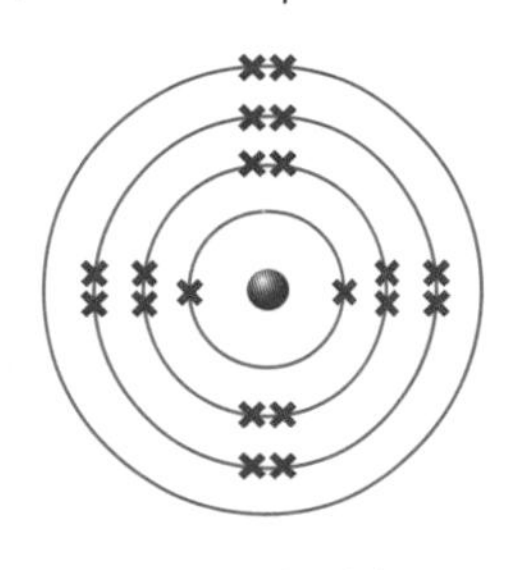

an atom of calcium Ca (2,8,8,2)

Check your understanding

Tested

1 Explain why atoms have no overall electrical charge. *(1 mark)*

2 The element tantalum (Ta) has an atomic number of 73 and a mass number of 181.
How many electrons, protons and neutrons are there in an atom of tantalum? *(3 marks)*

3 Draw diagrams to show the electronic structures of the following elements:
a) lithium *(1 mark)*
b) aluminium *(1 mark)*
c) phosphorus *(1 mark)*
d) potassium *(1 mark)*

4 Use symbols and numbers to show the electronic structures of the following elements:
a) beryllium *(1 mark)*
b) oxygen *(1 mark)*
c) magnesium *(1 mark)*
d) argon *(1 mark)*

5 How many electrons are there in the outer shells of the following atoms?
a) potassium and rubidium *(1 mark)*
b) aluminium and indium *(1 mark)*
c) calcium and barium *(1 mark)*
d) sulfur and polonium *(1 mark)*
e) fluorine and iodine *(1 mark)*

Answers online **Test yourself online** Online

The periodic table

Elements in the same **group** in the **periodic table** have the same number of **electrons** in their highest **energy level**. This gives them similar chemical properties.

Group 1

Revised

- Group 1 elements have **one electron** in their **outer shell** and react by losing this electron.

Reaction with water

- Group 1 elements react vigorously with water, dissolving and fizzing on the surface.

lithium + water → lithium hydroxide + hydrogen

$2Li(s) + 2H_2O(l) \rightarrow 2LiOH(aq) + H_2(g)$

- Sodium and potassium react in the same way:

sodium + water → sodium hydroxide + hydrogen

$2Na(s) + 2H_2O(l) \rightarrow 2NaOH(aq) + H_2(g)$

Reaction with oxygen

- Group 1 elements burn vigorously when heated in oxygen.
- Lithium and sodium react with oxygen in the same way as potassium:

potassium + oxygen → potassium oxide

$4K + O_2 \rightarrow 2K_2O$

The alkali metals
Lithium Li 2,1
Sodium Na 2,8,1
Potassium K 2,8,8,1

← The electron structures of the first three alkali metals

Group 0

Revised

- The elements in **group 0** are **the noble gases**.
- The noble gases are **unreactive** because their atoms have **stable** arrangements of electrons.
- This is because they have **full outer energy levels**.
- Helium **has two outer electrons**.
- All **other noble gases have eight outer electrons**.
- The noble gases have similar properties because they are in the same **group**.

The noble gases
Helium He 2
Neon Ne 2,8
Argon Ar 2,8,8

↑ The electron structures of the first three noble gases

Chemical reactions

- When elements react, their atoms join with other atoms to form **compounds**. This involves giving, taking or sharing **electrons** to form **ions** or **molecules**.
- When atoms **share** pairs of **electrons**, they form **covalent** bonds in **molecules**.

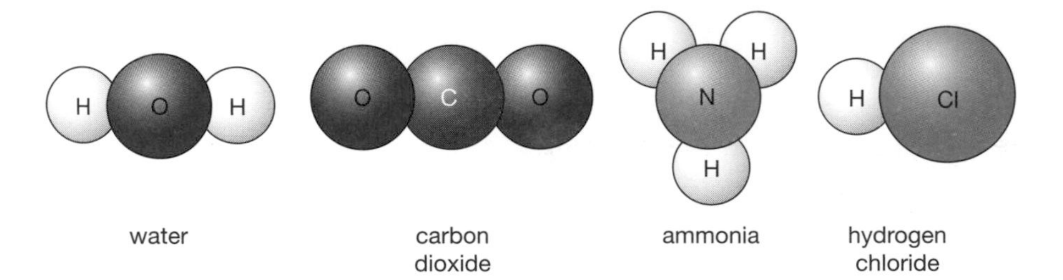

← Molecules of water, carbon dioxide, ammonia and hydrogen chloride

- Compounds formed from non-metals consist of **molecules**.
- In **molecules** the atoms are held together by **covalent bonds**.
- When atoms form chemical bonds by **transferring electrons**, they form **ions**.
- Compounds formed from metals and non-metals consist of **ions**.

- Metal atoms lose **electrons** to become positive ions.
- Non-metal atoms gain **electrons** to become negative ions.

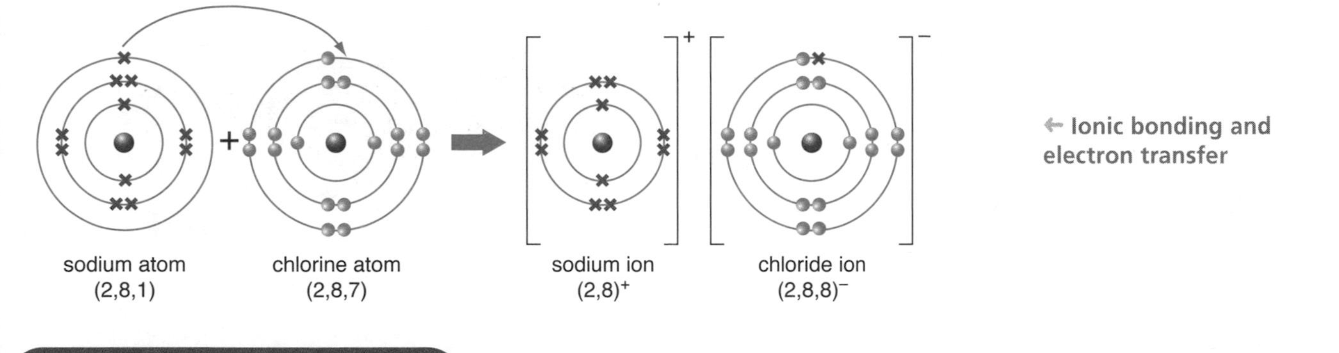

← Ionic bonding and electron transfer

Chemical equations

Revised

To write a **balanced chemical equation** there are four steps:

1. Write the **word equation**.
2. Rewrite the word equation using the **correct** chemical formulae.
3. **Balance** the numbers of atoms of reactants (left-hand side) with the number of product atoms (right-hand side).
4. Write out the balanced equation.

Example: Consider the reaction between magnesium and oxygen:

Step 1 magnesium + oxygen → magnesium oxide

Step 2 $Mg + O_2 \rightarrow MgO$

Step 3 On the left-hand side there are 1 magnesium and 2 oxygens
On the right-hand side there are 1 magnesium and 1 oxygen
The right-hand side is short of 1 oxygen and so we rewrite as:
$Mg + O_2 \rightarrow 2MgO$

The equation now balances for oxygen but not magnesium, so we have more to do:

Step 4 $\mathbf{2Mg + O_2 \rightarrow 2MgO}$ **Balanced!**

Since no atoms are lost or made in the reaction the **mass of the products** equals the **mass of the reactants**. So, in the above equation, the mass of $\mathbf{2Mg + O_2}$ = the mass of **2MgO**.

> **examiner tip**
> Never change a correct formula! In the example MgO was NOT changed into MgO_2.

> **examiner tip**
> Learn this method and make sure you use subscripts where necessary, or else you will lose marks.

Check your understanding

Tested

6 Which parts of an atom are involved in chemical bonding? *(1 mark)*

7 For the reaction between potassium and water, write:

a) the word equation *(1 mark)*

b) the balanced chemical equation *(2 marks)*

8 For the reaction between lithium and oxygen, write:

a) the word equation *(1 mark)*

b) the balanced chemical equation *(2 marks)*

9 The following equations are not balanced. Write them out and balance them.

a) $CH_4 + O_2 \rightarrow CO_2 + H_2O$ *(2 marks)*

b) $Na + H_2O \rightarrow H_2 + NaOH$ *(2 marks)*

c) $H_2SO_4 + NaOH \rightarrow Na_2SO_4 + H_2O$ *(2 marks)*

10 What mass of oxygen reacts with 48 g of magnesium to produce 80 g of magnesium oxide? *(1 mark)*

Answers online **Test yourself online** Online

Calcium carbonate

- Rocks provide essential building materials.
- Limestone is a naturally occurring resource that provides a starting point for the manufacture of cement and concrete.
- Limestone, which is mainly calcium carbonate ($CaCO_3$), is quarried and can be used as a building material.
- Calcium carbonate can be **thermally decomposed** by strong heating to make calcium oxide and carbon dioxide:

$CaCO_3 \rightarrow CaO + CO_2$

- The carbonates of magnesium, copper, zinc and sodium **thermally decompose** in the same way, for example:

$Na_2CO_3 \rightarrow Na_2O + CO_2$

- Some **group 1 carbonates** cannot be decomposed using a Bunsen burner — it cannot get hot enough.

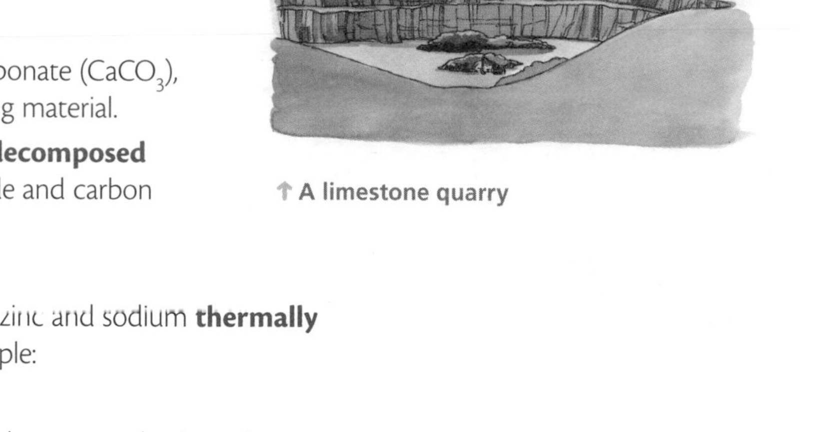

↑ A limestone quarry

The limestone cycle

Revised

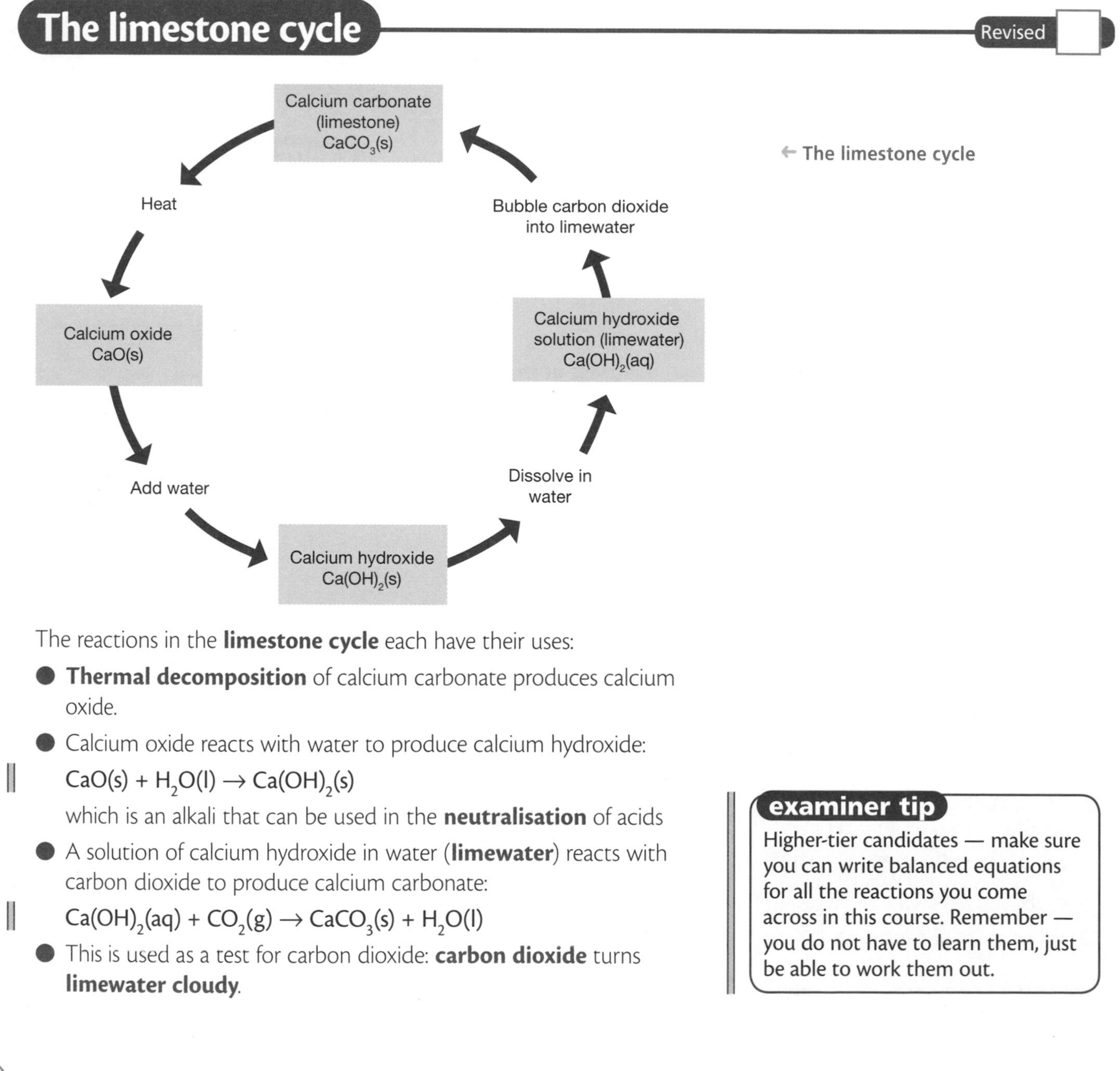

← The limestone cycle

The reactions in the **limestone cycle** each have their uses:

- **Thermal decomposition** of calcium carbonate produces calcium oxide.
- Calcium oxide reacts with water to produce calcium hydroxide:

$CaO(s) + H_2O(l) \rightarrow Ca(OH)_2(s)$

which is an alkali that can be used in the **neutralisation** of acids

- A solution of calcium hydroxide in water (**limewater**) reacts with carbon dioxide to produce calcium carbonate:

$Ca(OH)_2(aq) + CO_2(g) \rightarrow CaCO_3(s) + H_2O(l)$

- This is used as a test for carbon dioxide: **carbon dioxide** turns **limewater cloudy**.

examiner tip

Higher-tier candidates — make sure you can write balanced equations for all the reactions you come across in this course. Remember — you do not have to learn them, just be able to work them out.

- Carbonates react with acids to produce carbon dioxide, a salt and water, for example:

copper carbonate + nitric acid → carbon dioxide + copper nitrate + water

$$CuCO_3 + 2HNO_3 \rightarrow CO_2 + Cu(NO_3)_2 + H_2O$$

examiner tip

Remember, limestone is a carbonate and so statues and buildings made from it are damaged by acid rain.

Making limestone useful

Revised

To do anything with limestone, other than use it as the rock, requires chemistry to modify its properties.

- **Building blocks** — quarried and cut to shape.
- **Cement** — heat limestone with clay.
- **Mortar** — mix cement with sand and water.
- **Concrete** — mix mortar with small stones (aggregate).

examiner tip

Limestone is used as building blocks, but it is used to make the other materials. Take care with your wording.

Advantages and disadvantages of using these products

Revised

Here are some factors about limestone, cement and concrete to consider:

- Limestone buildings are **good to look at**.
- Limestone is fairly **cheap** and so are its products.
- Concrete can be **poured** into moulds to make different shapes.
- Limestone, concrete, and bricks and mortar are **stronger** than timber buildings.

Here are some factors about quarrying to consider:

- Quarrying provides **employment**.
- The hole is **ugly**.
- The quarry provides **necessary rock** for producing building materials.
- There has been a loss of wildlife **habitat**.
- The lorries are **dirty**.
- The blasting is **noisy**.
- The works produce a lot of **smoke pollution**.

examiner tip

If you are asked to **evaluate** the information supplied, you must remember to give a **balanced** argument and deal with the **evidence for and against** — forget one and you will lose marks.

Check your understanding

Tested

11 For the thermal decomposition of zinc carbonate, write:

a) the word equation *(2 marks)*

b) the balanced chemical equation *(1 mark)*

12 For the reaction of water with calcium oxide, write:

a) the word equation *(2 marks)*

b) the balanced chemical equation *(2 marks)*

13 For the reaction of magnesium carbonate with hydrochloric acid, write:

a) the word equation *(2 marks)*

b) the balanced chemical equation *(2 marks)*

14 List two advantages and two disadvantages of using limestone as a building material. *(4 marks)*

Answers online **Test yourself online** Online

Extracting metals

- Metals are very useful in our everyday lives.
- Ores are naturally occurring rocks that provide an economic starting point for the manufacture of metals.
- Iron ore is used to make iron and steel.
- Copper can be easily extracted but copper-rich ores are becoming scarce so new methods of extracting copper are being developed.
- Aluminium and titanium are useful metals but are expensive to produce.
- **Ores** contain enough metal to make it **economical** to extract the metal.
- The economics of extraction may **change** over time. If the price of the metal goes down, it may not be economic to extract it from a **low-grade** ore.

Earth's crust

Rocks – some rocks are **ores**

metals, e.g. gold

metals compounds in minerals

always mixed with other substances

no chemicals needed for extraction – just crush and separate from waste rock

chemical reaction needed for extraction – e.g. removing oxygen by reduction

↑ **Extracting metals and metal compounds from the Earth's crust**

You should be able to analyse data like these to explain how decisions about whether to mine metal ores depend on the concentration of metal and on the price:

Percentage of tin in ore	Price of tin per tonne	Income per tonne of tin produced	Mining cost per tonne of ore	Profit per tonne of tin produced
2.5	£5 500	£138	£100	£38
1.5	£7 000	£105	£100	£5
1.0	£6 000	£60	£100	£40
1.0	£10 000	£100	£100	£0

- Ores are mined and may be **concentrated** before the metal is extracted and purified.
- **Unreactive** metals such as **gold** are found in the Earth as the **metal** itself.
- **More reactive** metals are found as minerals in **ores**.
- Minerals are **compounds** – a **chemical reaction** is needed to extract the metal.

Reduction using carbon

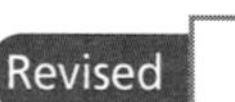

- Reduction is the removal of oxygen.
- Metals that are **less reactive** than carbon can be extracted from their oxides by **reduction** with carbon, for example iron oxide is reduced in the blast furnace to make iron:

  iron oxide + carbon monoxide → iron + carbon dioxide

  $Fe_2O_3 + 3CO \rightarrow 2Fe + 3CO_2$

- Copper can be extracted from copper-rich ores by heating the ores in a furnace (smelting):

  copper oxide + carbon monoxide → copper + carbon dioxide

  $CuO + CO \rightarrow Cu + CO_2$

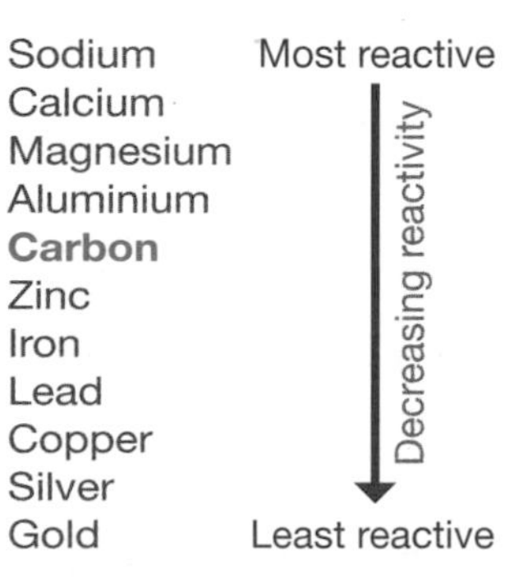

↑ **The reactivity series – note that aluminium is above carbon, but iron is below carbon**

examiner tip

When carbon is used in **reduction**, carbon monoxide (CO) is often the **reducing agent** in the equation.

Copper: the useful metal

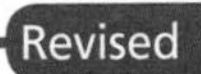

- The supply of copper-rich ores is limited and is running out.
- Traditional mining and extraction have major **environmental impacts**.
- New ways of extracting copper from **low-grade** ores are being researched to limit the environmental impact of traditional mining.
- Copper can be extracted by **phytomining**, or by **bioleaching**.

Phytomining uses plants to **absorb** metal compounds. The plants are burned to produce ash that contains the metal compounds.

- **Bioleaching** uses bacteria to produce **leachate solutions** that contain metal compounds.

Electrolysis

Revised

- Metals that are **more reactive** than carbon, such as **aluminium**, are extracted by **electrolysis** of molten compounds.
- The use of large amounts of **energy** in the extraction of these metals makes them **expensive**.
- The **copper** from smelting can be purified by **electrolysis**.
- **Copper** can be obtained from solutions of copper salts by **electrolysis** or by **displacement** using scrap iron, for example in the displacement of copper from copper sulfate solution:

copper sulfate + iron → copper + iron sulfate

$CuSO_4(aq) + Fe(s) \rightarrow Cu(s) + FeSO_4(aq)$

- During electrolysis **positive ions** move towards the **negative electrode**.
- **Aluminium and titanium** cannot be extracted from their oxides by reduction with carbon. Current methods of extraction are **expensive** because:
  - there are many **stages** in the processes
  - large amounts of **energy** are needed

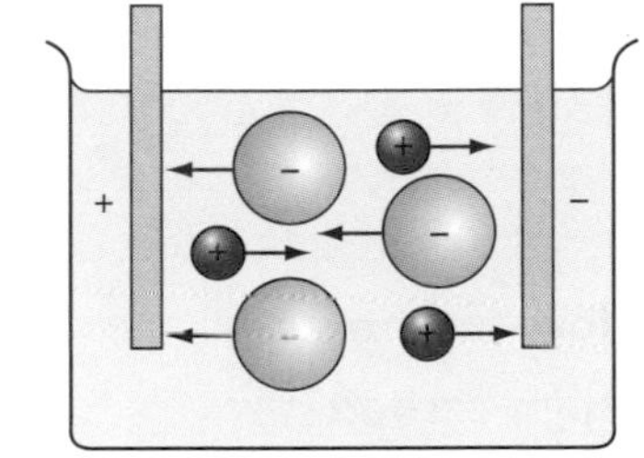

↑ Positive ions move towards the negative electrode

Why recycle metals?

Revised

- **It is cheaper** in terms of **energy** to recycle than to dig out more ore.
- **It is cheaper** in terms of **the environment** — no mining means fewer spoil heaps, and there are virtually no emissions of harmful gases (some methods of processing copper ore do release sulfur dioxide).
- With **limited resources**, it makes sense to conserve remaining ore.

Check your understanding

Tested

15 Which metals are found as themselves in the Earth — the least reactive or most reactive? *(1 mark)*

16 By what process is iron extracted from its ore? *(1 mark)*

17 Why can titanium not be extracted by reduction with carbon? *(1 mark)*

18 Why is aluminium an expensive metal? *(2 marks)*

19 Why should we make the effort to recycle metals? *(3 marks)*

Answers online **Test yourself online** Online

Alloys

Metals can be mixed together to make alloys.

Iron and steel

Revised

- Iron from the blast furnace (sometimes used as **cast iron**) contains about 96% iron.
- The impurities make it **brittle**, so cast iron has limited uses but it is strong in compression.
- Most iron is converted into **steels**.
- Steels are **alloys** — they are **mixtures** of iron, carbon and, sometimes, other metals.

Properties of alloys

Revised

Alloys can be designed to have properties for specific uses (by varying the amounts of other metals that are added).

In the case of steel, it is the amount of carbon added that has the greatest effect on the properties:

- **low-carbon** steels are easily shaped
- **high-carbon** steels are hard
- **stainless** steels are resistant to corrosion

examiner tip

If the question asks you to 'give a reason', you need to apply scientific knowledge, but you can use short sentences or bullet points.

Make sure you say 'lightweight', not just 'light'. After all, a tonne of steel weighs the same as a tonne of aluminium. Even better, say 'low-density'.

Everyday metals

Revised

Many metals in everyday use are alloys. Pure copper, gold, iron and aluminium are too soft for many uses and so are mixed with small amounts of similar metals to make them harder for everyday use:

- Stainless steel is used to make cutlery.
- Coins are made of various alloys of copper and nickel.
- Brass door handles are an alloy of copper and zinc.
- Pure gold rings would be far too soft to keep their shape, so copper is added to increase their strength.
- Aluminium is often alloyed with magnesium to make it strong enough for everyday use.

Properties and uses of metals

The elements in the central block of the periodic table are known as **transition metals**. The **transition metals**, like other metals, are good **conductors** of heat and electricity, and can be bent or hammered into shape. They are useful as structural materials and for making things that must allow heat or electricity to pass through them easily.

He
TRANSITION METALS
NON-METALS
Ti Cr Mn Fe Ni Cu Zn
REACTIVE METALS
Ag
POOR METALS
Pt Au

← Transition metals are the central block of the periodic table

Copper is useful for **electrical wiring** and **plumbing**:

- It is a good **conductor** of **electricity** and **heat**.
- It can be **bent** but is **hard** enough to be used to make pipes or tanks.
- It **does not react** with **water**.

Aluminium and titanium

Revised

- **Low density** makes aluminium and titanium useful. The lightest bikes and the fastest jets are made from aluminium and titanium.
- Their **resistance to corrosion** is also useful. Aluminium window frames resist corrosion because of the tough oxide layer on them — so they do not need painting.

↑ Aluminium and titanium are used to make aircraft because they resist corrosion and have low densities

Benefits and drawbacks of using metals

Revised

- Iron from the blast furnace is brittle, but strong under compression. It is good for manhole covers, but dangerous for building bridges and girders.
- High-carbon steel is also brittle, so it is not suitable for use as steel girders, but it is good for drill bits — they need to be hard and stay sharp; low-carbon steel would be too soft.
- Copper and aluminium are soft, so their harder alloys are used, rather than the pure metals — you cannot make an aircraft of pure aluminium.

examiner tip

In the exam, you may have to evaluate the benefits, drawbacks and risks of using different metals. You need to consider the metal's properties.

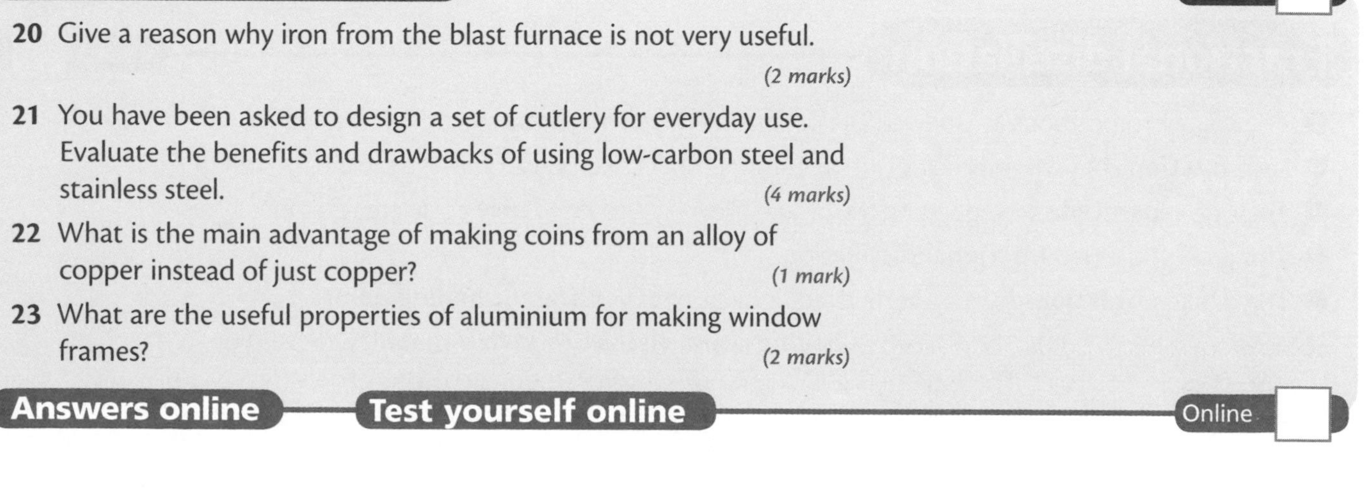

Check your understanding

Tested

20 Give a reason why iron from the blast furnace is not very useful. *(2 marks)*

21 You have been asked to design a set of cutlery for everyday use. Evaluate the benefits and drawbacks of using low-carbon steel and stainless steel. *(4 marks)*

22 What is the main advantage of making coins from an alloy of copper instead of just copper? *(1 mark)*

23 What are the useful properties of aluminium for making window frames? *(2 marks)*

Answers online **Test yourself online** Online

Crude oil

- Crude oil is derived from ancient biomass found in rocks.
- Many useful materials can be produced from crude oil.
- Crude oil can be fractionally distilled.
- Crude oil is a **mixture** of a very large number of compounds.
- A **mixture** consists of two or more elements or compounds **not chemically combined** together.
- The chemical properties of each substance in the mixture are unchanged.
- It is possible to **separate** the substances in a mixture by physical methods, including **distillation**.

Hydrocarbons

- Most of the compounds in crude oil consist of molecules made up of hydrogen and carbon atoms only.
- **Hydrocarbons** are molecules made up of hydrogen and carbon atoms **only**.
- Most of these are **saturated** hydrocarbons called **alkanes**, which have the general formula $\mathbf{C_nH_{2n+2}}$.
- **Saturated** means you cannot fit any more hydrogen atoms into the molecule because there are no **carbon–carbon double bonds**.
- **Alkane** molecules can be represented in the forms shown in the diagrams below. Can you see that no more hydrogen atoms will fit into the **alkanes**?
- The names and formulae of the **first four alkanes** are:
  - **methane**, CH_4
  - **ethane**, C_2H_6
  - **propane**, C_3H_8
  - **butane**, C_4H_{10}
- In these **displayed structures** below, each single line represents a **covalent bond**.

examiner tip

For top marks learn the definition of a hydrocarbon, **word for word**.

examiner tip

Make sure you can recognise an alkane from any form of its formula.

Learn the **names** and **formulae** of the first four alkanes.

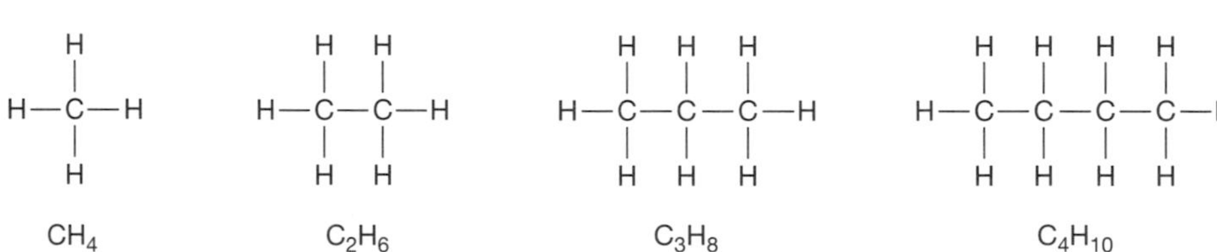

Fractional distillation

Revised

- The many hydrocarbons in crude oil can be separated into **fractions**.
- Each **fraction** contains molecules with a similar number of carbon atoms (see table opposite).
- They are **separated** by evaporating the oil and allowing it to **condense** at different temperatures.
- This process is called **fractional distillation**.
- The different **fractions** are easily separated because of their **different boiling points**.

Some properties of hydrocarbons, such as **boiling point**, **viscosity** and **flammability**, depend on the size of their molecules. These properties influence how hydrocarbons are used as **fuels** (see pages 16–17).

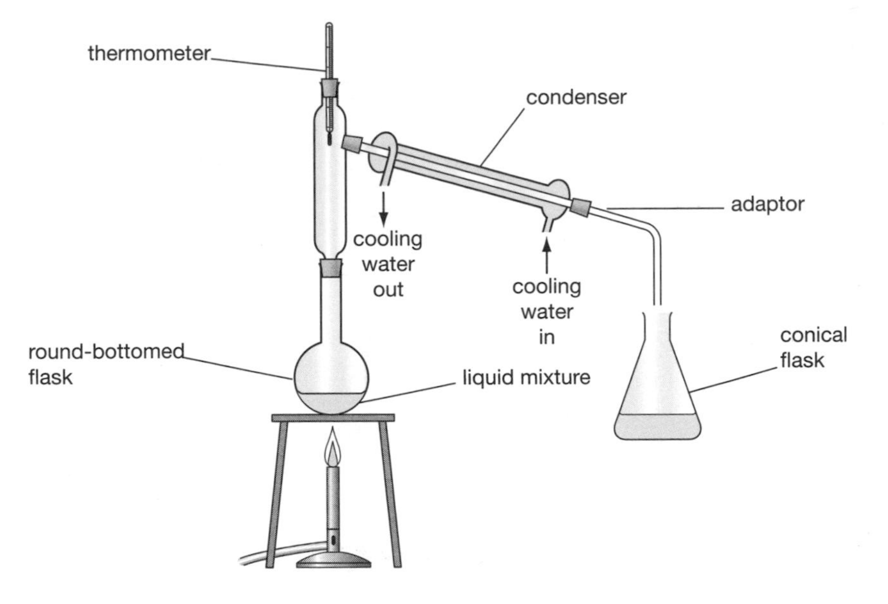

← Fractional distillation in the laboratory — the mixture is heated and the conical flask is changed as the heated vapour condenses at increasing temperatures

Note these trends from the table. As the molecules get bigger:

- the boiling points increase
- the viscosity increases
- the flammability becomes less

This could lead to a **hypothesis**, such as boiling point is related to molecule size — a graph would show the **correlation** better.

examiner tip

You do not need to learn the names in the table, but you need to know the trends in the boiling points, viscosity and flammability of the fractions as the molecules get bigger.

Fractions from an oil refinery

Fraction	Number of carbon atoms in molecule	Description and viscosity	Flammability	Boiling point (°C)	Uses
Refinery gas	1–4	Colourless gases	Explodes if mixed with air and lit	Less than 40	Used as a fuel in the refinery; bottled and sold as liquefied petroleum gas
Naphtha	5–10	Yellowish liquid, flows very easily	Evaporates easily, vapour mixed with air is explosive	25–175	Petrol
Kerosene	10–14	Yellowish liquid, flows like water	Will burn when heated	150–260	Aircraft fuel
Light gas oil	14–20	Yellow liquid, thicker than water	Needs soaking onto a wick or other material to burn	235–360	Diesel fuel
Heavy gas oil	20–50	Yellow-brown liquid	Just burns when soaked onto a wick — very smoky	330–380	Used in catalytic crackers (see page 18)
Fuel oil	60–80	Thick, brown, sticky liquid	Needs to be hot and soaked onto a wick before it will burn	Above 490	Fuel oil for power stations and ships

Check your understanding

Tested

24 Why is air a mixture, not a compound? *(1 mark)*

25 What is similar about the molecules in a crude oil fraction? *(2 marks)*

26 Why is C_2H_5OH not a hydrocarbon but $C_{10}H_{22}$ is? *(2 marks)*

27 What properties of crude oil fractions are related to molecule size, and how do they change as the molecules become larger? *(4 marks)*

Answers online **Test yourself online** Online

Hydrocarbon fuels

Products of combustion

Revised

- Most fuels, including coal, contain **carbon** and/or **hydrogen**.
- Some fuels also contain **sulfur**.
- So the gases released into the atmosphere when a fuel burns may include **carbon dioxide**, **water** (vapour), **carbon monoxide**, **sulfur dioxide** and **oxides of nitrogen**.
- Solid particles (**particulates**) such as **soot** (carbon) or **unburnt hydrocarbons** may also be released.
- The combustion of hydrocarbon fuels **releases energy**.
- During combustion the carbon and hydrogen in the fuels are **oxidised**.
- The equation for burning methane is:

methane + oxygen → carbon dioxide + water

$$CH_4 + 2O_2 \rightarrow CO_2 + 2H_2O$$

examiner tip

Make sure you can balance the equation for alkanes up to at least butane (C_4H_{10}).

Harmful emissions from burning fuels

Revised

- **Sulfur dioxide** and **oxides of nitrogen** cause **acid rain** — they dissolve in the water to produce sulfuric acid and nitric acid.
- **Carbon dioxide** emissions cause **global warming** — due to an increased greenhouse effect.
- **Solid particles** cause **global dimming** — sunlight cannot get through the atmosphere as easily, and atmospheric temperature falls.
- **Incomplete combustion** can also produce **carbon monoxide**, which is poisonous.
- **Oxides of nitrogen** are formed at high temperatures, for example in a car engine.

Impact on the environment

Revised

What happens when we burn fossil fuels? The diagram says it all.

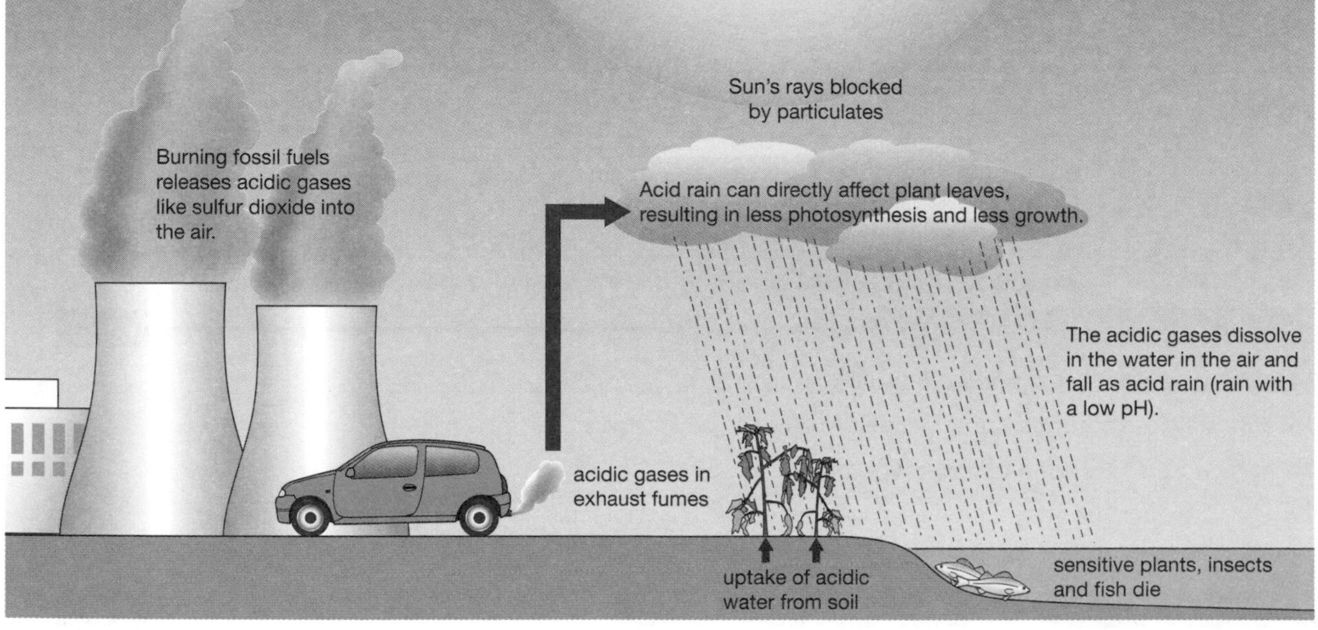

↑ The environmental impacts of burning fossil fuels

Cleaner fuels

Revised

- Sulfur in fuel eventually causes **acid rain**.
- Sulfur can be removed from fuels **before** they are burned, for example in vehicles.
- In power stations, sulfur dioxide can be removed from the waste gases **after** combustion.

Alternative fuels

Revised

- Biodiesel and ethanol are produced from plant material. There are economic, ethical and environmental issues surrounding their use.
- Sugar cane **absorbs** carbon dioxide as it grows. When the ethanol produced from it burns, no **new** carbon dioxide is released.
- **Hydrogen** is made from water, and can be used in an engine or in **fuel cells** to produce electricity.
- Ethanol and hydrogen have advantages over hydrocarbons as fuels:
  - They do not deplete **non-renewable** crude oil reserves.
  - They are 'clean' — they produce no toxic fumes, and using hydrogen does not produce carbon dioxide.

The drawback of **ethanol** is that it takes a lot of sugar cane to make the fuel — the fields could be used to produce food.

There is a problem with **storing** hydrogen as it is a gas and takes up a large volume. It needs to be pressurised or liquefied, which takes a lot of energy.

examiner tip

If you are asked for economic and environmental effects of using fuels, you must remember to give a balanced argument and deal with all the aspects.

Energy outputs

Revised

Some cleaner fuels do not give out as much energy as the dirty ones. Decide which of these you would put in your petrol tank. Why?

	Diesel	Biodiesel	Ethanol	Hydrogen (liquid)
Energy content per gallon (British thermal units)	130 000	120 000	80 000	30 500

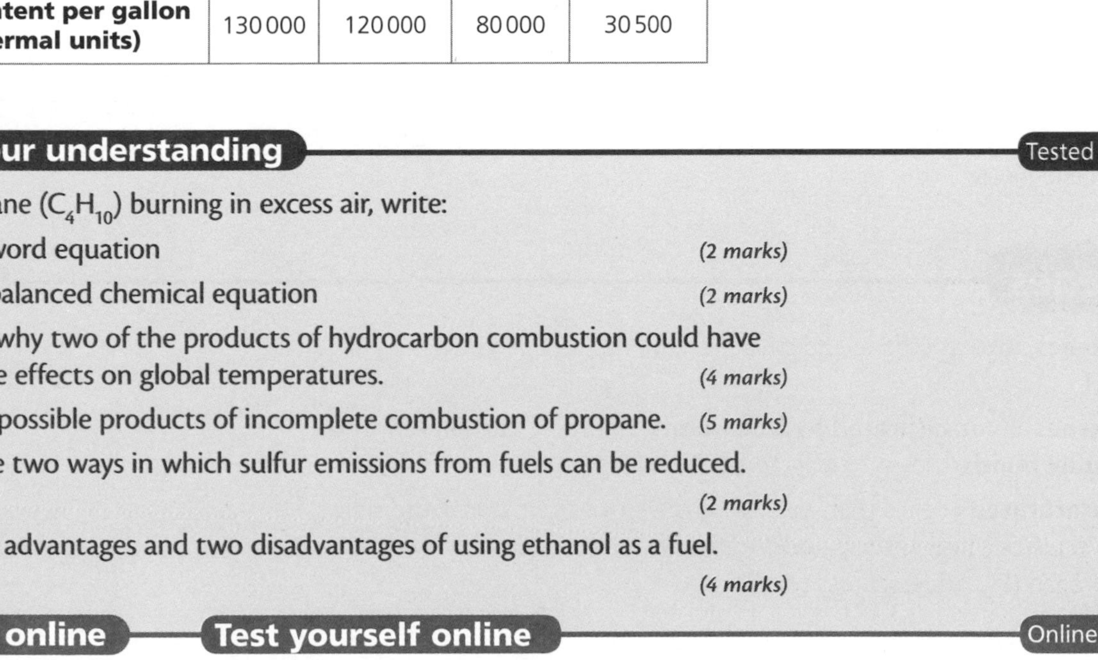

Check your understanding

Tested

28 For butane (C_4H_{10}) burning in excess air, write:

a) the word equation *(2 marks)*

b) the balanced chemical equation *(2 marks)*

29 Explain why two of the products of hydrocarbon combustion could have opposite effects on global temperatures. *(4 marks)*

30 List the possible products of incomplete combustion of propane. *(5 marks)*

31 Describe two ways in which sulfur emissions from fuels can be reduced. *(2 marks)*

32 List two advantages and two disadvantages of using ethanol as a fuel. *(4 marks)*

Answers online **Test yourself online**

Online

Obtaining useful substances from crude oil

Fractions from the distillation of crude oil can be broken down (cracked) to make smaller molecules including unsaturated hydrocarbons such as ethene.

Cracking crude oil

Revised

- Hydrocarbons can be broken down (**cracked**) to produce smaller, more useful molecules.
- **Cracking** involves heating the hydrocarbons to vaporise them.
- The vapours are either passed over a hot **catalyst** or mixed with **steam** and heated to a very high temperature.
- A catalyst speeds up a reaction but is not consumed (not used up by the reaction).
- Cracking involves **thermal decomposition** reactions — this is using heat to break the molecules apart.

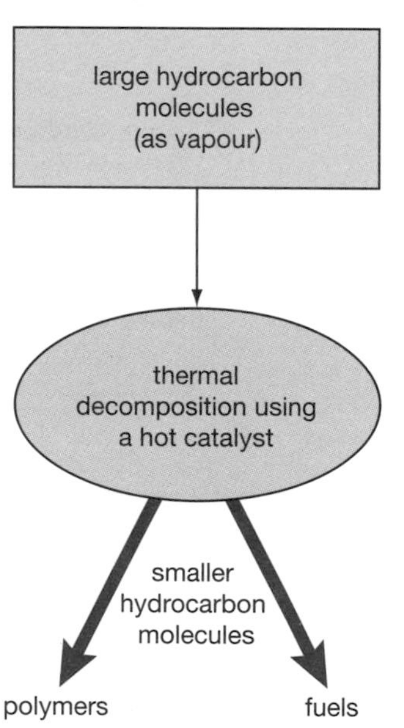

Cracking is the process that breaks large hydrocarbon molecules into smaller ones

The products of cracking alkanes include:

- **alkanes** that are useful as **fuels** (they are shorter than the original alkanes)
- **alkenes**, which are also used as fuels and as raw materials for plastics and chemicals

Alkenes

Revised

- **Alkenes**, such as ethene and propene, have the general formula C_nH_{2n}.
- **Alkenes** are **unsaturated hydrocarbons** containing **carbon–carbon double bonds**, shown in diagrams by C=C.
- '**Unsaturated**' means that you can fit more hydrogen atoms into the molecule because of the carbon–carbon double bonds, which can have hydrogen (H_2) added across them.

- Unsaturated hydrocarbon molecules can be represented in the following forms:

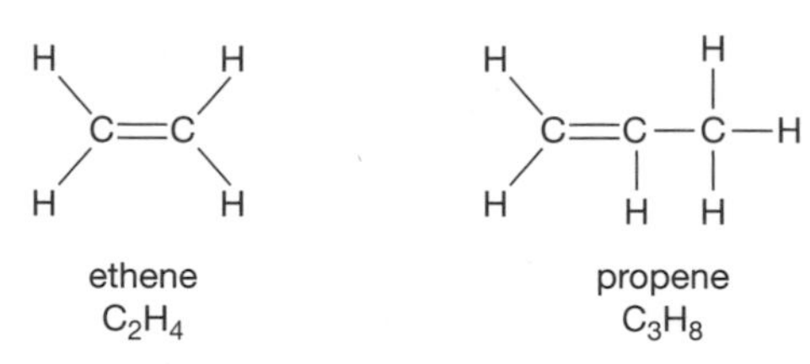

- In the **displayed structures** of alkenes each double line (═) represents a **double bond**.
- It is the C═C double bond that makes alkenes reactive and useful. The C═C bond also makes alkenes easy to detect — even in substances like margarine.

> **examiner tip**
> Make sure you can recognise an alkene from its name or formula. You only need to **learn** the **names** of the first two alkenes — **ethene** and **propene**.

Testing for alkenes

Revised

- Alkenes can be detected using **bromine water**.
- It is a test for **unsaturation** in a hydrocarbon.
- You may have seen this with ethene gas (when you cracked an **alkane** like paraffin) and/or with a vegetable oil like margarine (messy!). (See page 28.)
- To test for an alkene, you shake some bromine water with the suspect material. If it does contain an alkene then the **bromine water** turns from **orange** to **colourless** (it is decolourised).
- The bromine water loses its distinctive colour because it reacts with the double bonds, adding across them, just as hydrogen does. (See page 28.)

> **examiner tip**
> When describing a colour change, make sure you say what the colour was to start with and what it was at the end. Bromine water **starts orange** and **becomes colourless**.

Check your understanding

Tested

33 What is a catalyst? *(2 marks)*.

34 What type of reaction is involved in cracking hydrocarbons? *(1 mark)*

35 Describe how margarine will react with bromine water. *(2 marks)*

36 Draw a diagram of an ethene molecule. Make sure you show **all** the bonds. *(2 marks)*

Answers online **Test yourself online** Online

Polymers: production and uses

- Unsaturated hydrocarbons can be used to make polymers.
- **Polymers** are extremely long molecules or chain molecules.
- **Alkenes** can be used to make **polymers** such as **poly(ethene)** and **poly(propene)**.
- In **polymerisation** reactions, many small molecules (**monomers**) join together to form very large molecules (**polymers**). For example, ethene molecules join up to make **poly(ethene)** and propene molecules join up to make **poly(propene)**.

A polymer chain can be made up of between 10 000 and 50 000 monomer molecules all joined together. If you made a model of such a polymer chain using the standard laboratory model kits you would have to make it at least 500 metres long. 2500 metres would not be too long either!

There are two ways in which you can represent the polymerisation of a monomer. Like this:

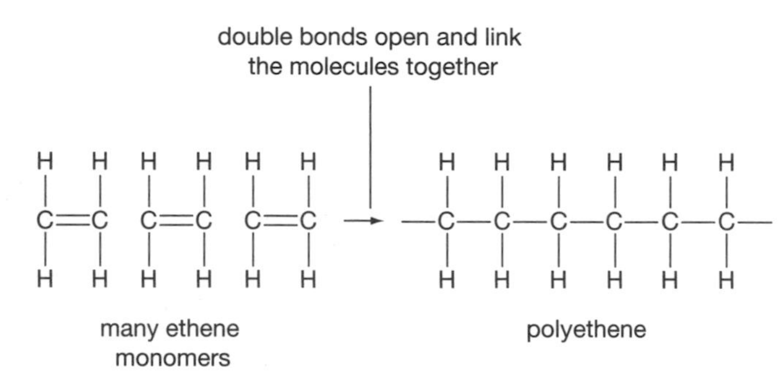

↑ **A short length of poly(ethene), which is a long-chain polymer molecule**

Or like this:

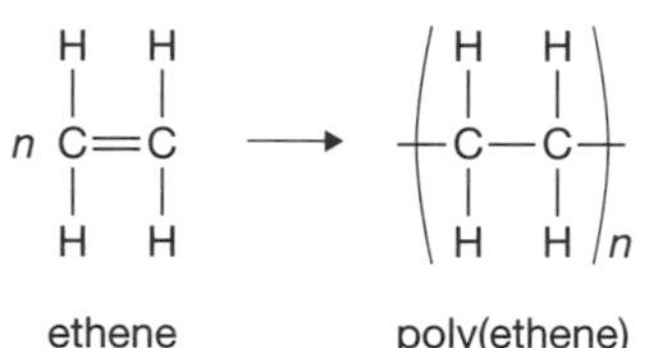

↑ **An alternative way of representing the formation of poly(ethene); *n* is a large number**

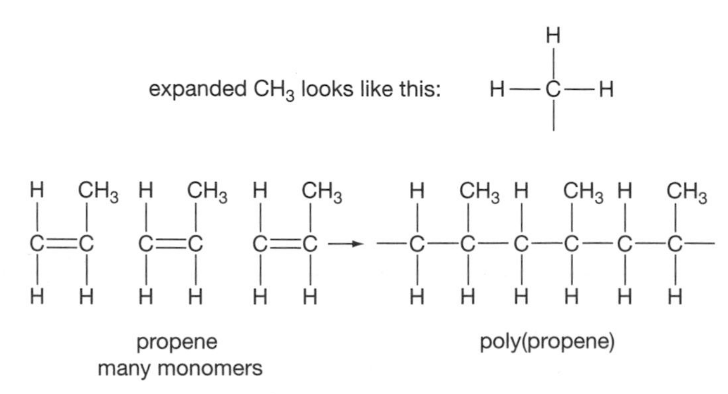

↑ **A short length of poly(propene), which is another long-chain polymer molecule**

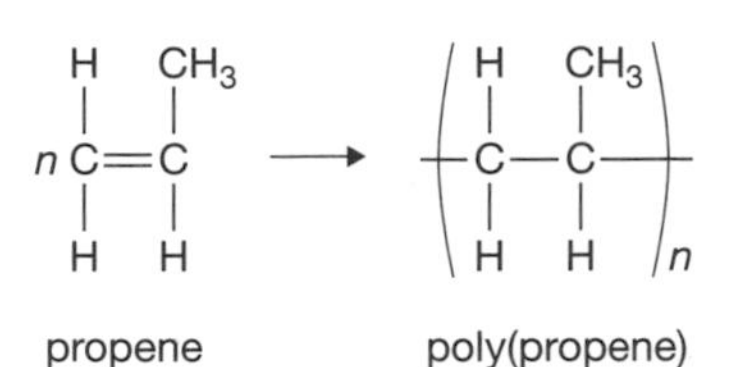

↑ **An alternative way of representing the formation of poly(propene)**

examiner tip

Make sure you can represent the formation of a polymer from a given alkene monomer. Also make sure you draw your bonds **accurately**, they must go to the **carbon** of the CH_3 groups.

Designing different polymers

Revised

Polymers have many useful applications and new uses are being developed all the time.

Here are some examples of new polymers made with special properties:

- **New packaging materials** — smart packaging materials can change colour when the temperature changes or the sell-by date is reached.
- **Waterproof coatings** for fabrics such as Gore-Tex®, which are actually layers of polymers designed with tiny holes that let through water vapour (perspiration), but not liquid water (rain).
- **Dental polymers** — tooth-coloured fillings are moulded into shape and then hardened by exposure to a specific blue light.
- **Hydrogels** are polymer gels that grow or shrink when water is added — they are used to mop up chemical or oil spills and in nappies to keep the baby's skin dry.
- **Wound dressings** can include special hydrogels that can absorb the discharge from wounds and improve the rate of healing.
- **Shape-memory polymers** can bend or stretch, but return to their original shape when heated — they are finding particular uses in surgery.

Crude oil: too valuable to burn?

Revised

Some people say it is. They believe that the limited supply of crude oil should be reserved for making **useful products** rather than being used as **energy sources**.

Can you evaluate (work out) the social and economic advantages and disadvantages of using products from **crude oil** as **fuels** or as **raw materials** for plastic and other chemicals? There is more about this in the next section.

Check your understanding

Tested

37 Many polymers are plastics. Name the plastic that can be made from propene. *(1 mark)*

38 Fill in the missing words: *(4 marks)*

Polymers can be made from ______**a**______ . In these reactions, many small molecules (______**b**______) join together to form very ______**c**______ molecules (polymers). For example, ______**d**______ molecules join up to make poly(ethene).

39 How could a shape-memory polymer be used as a surgical suture (stitch)? *(2 marks)*

40 Give a benefit and a drawback of using disposable nappies that contain hydrogel. *(2 marks)*

Answers online **Test yourself online** Online

Polymers: disposal

- Many polymers are **not biodegradable**, so they are not broken down (**decomposed**) by microbes, and this can lead to problems with waste disposal.
- **Biodegradable** plastics made from corn starch have been developed.
- Plastic bags are being made from biodegradable **polymers** so that they break down more easily.
- The problem is in developing a bag **to last as long** as it is needed before it **rots** away.

What happens to non-biodegradable polymer waste?

Revised

Polymer waste can be disposed of in three ways:

- in a landfill site
- by recycling
- by burning in a high-temperature incinerator

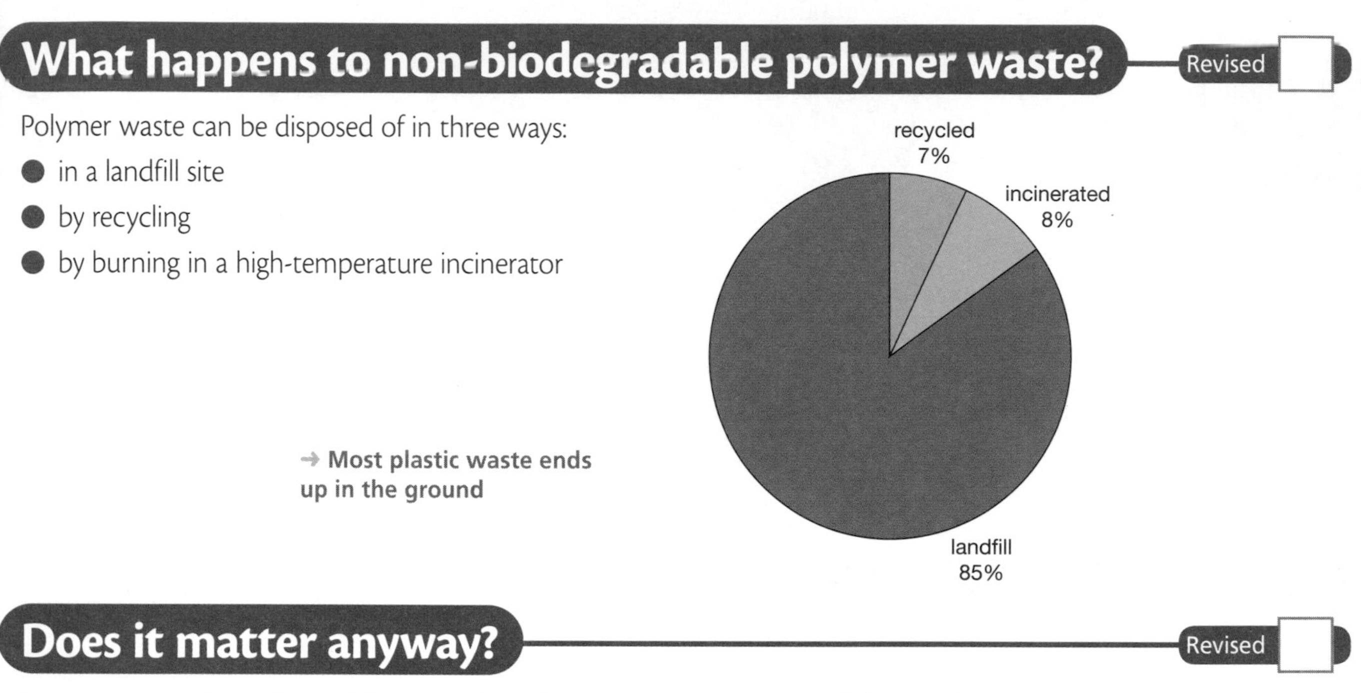

→ **Most plastic waste ends up in the ground**

Does it matter anyway?

Revised

Let us try to evaluate the social, economic and environmental impacts of the uses, disposal and recycling of polymers. Here is some information for you to consider:

- In New York City alone, one fewer bag per person per year would save \$250 000 in disposal costs.
- When 1 tonne of plastic bags is re-used or recycled, the energy equivalent of 11 barrels of oil is saved.
- Experts estimate that 500 billion to 1 trillion plastic bags are consumed and discarded annually worldwide — that is more than a million per minute.

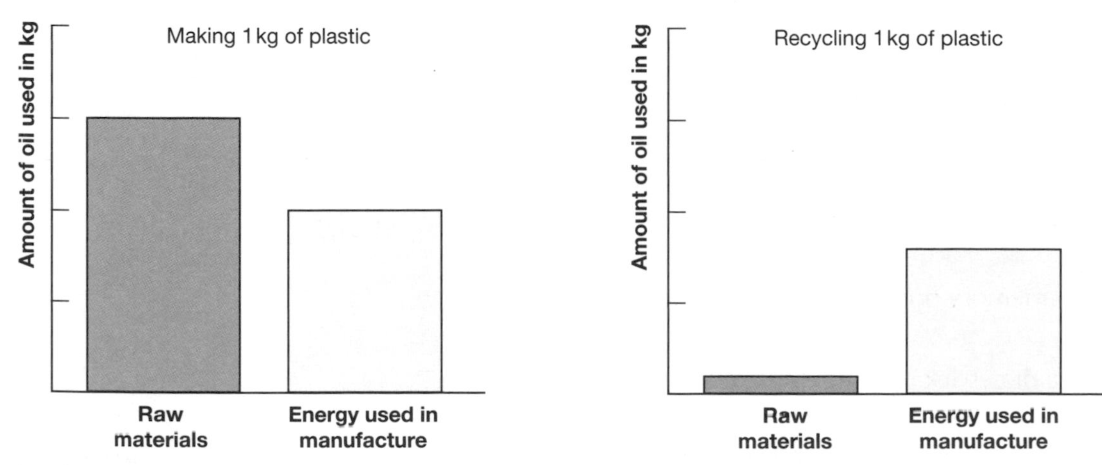

↑ **Recycling plastic saves oil resources, both as raw materials and as energy used in manufacturing**

Tables are useful tools when it comes to looking at lots of information together. The table below has a few ideas for you.

Tables are also useful when it comes to evaluation. If you use a table to do an evaluation for the three different impacts of recycling polymers it should look very similar to the one below.

Learn the following **evaluation** technique:

- What are the advantages and disadvantages (benefits and drawbacks)? You can make columns for these.
- Use the information provided, but check for the views of people who might be biased.
- Use rows in your table to separate the concepts, ideas or materials you are evaluating.
- Use the evidence of data in graphs or tables to support your argument.
- Come to your conclusion.

	Social	**Economic**	**Environmental**
Uses of polymers	Many products are made of polymers, e.g. sports equipment, clothing, food packaging, ICT and high-tech items…	Creates employment — manufacture, sales, disposal, product design…	Uses fossil fuel (oil) as raw material; manufacture may produce toxic waste; creates non-biodegradable waste at end of useful life
Disposal of polymers	Unpleasant to live close by	Creates employment — refuse collectors, landfill/ incinerator workers	Landfill uses valuable land and is unsightly; incineration produces a lot of ash, greenhouse gases and toxic fumes, but can also be used to generate energy
Recycling of polymers	Happier communities because their environment is cleaner, but there is effort involved in individuals having to sort domestic waste	Creates employment — is recycled plastic more expensive than new plastic?	Saves valuable raw materials; produces much less pollution; reduces use of landfill sites; uses energy — possibly more than the energy needed to make new plastic

examiner tip

If you are asked to **evaluate** the information supplied, you must remember to give a **balanced** argument and deal with the **evidence for and against** — forget one and you will lose marks.

Check your understanding

Tested

41 Give three benefits of recycling supermarket carrier bags — one social, one economic and one environmental. *(3 marks)*

42 Using the information on this page, evaluate the social and economic advantages and disadvantages of using products from crude oil as fuels, or as raw materials for plastics and other chemicals. *(6 marks)*

Answers online **Test yourself online** Online

Ethanol

We have already seen that alkane fuels are often thought to be 'dirty' — they produce pollutants. **Ethanol** is a much 'cleaner' fuel to use in car engines — it burns to produce **water** and **carbon dioxide** only. There are no particulates and no sulfur dioxide.

Ethanol from ethene

Revised

- **Ethene** is an alkene produced by cracking the heavier fractions in crude oil.
- **Ethanol** can be produced by **hydration** of **ethene** with steam in the presence of a catalyst.
- **Hydration** is simply the chemical addition of water to a molecule:

$$C_2H_4 + H_2O \xrightarrow[\text{high pressure}]{\text{catalyst}} C_2H_5OH$$

ethene + steam ⟶ ethanol

Only 5% of the ethene is converted, but by removing the ethanol from the mixture and recycling the ethene through the reactor, 95% conversion is eventually achieved.

The reaction has to be done under a pressure of about 300 atmospheres and the ethanol has to be separated from the water by fractional distillation, but the final product is very clean.

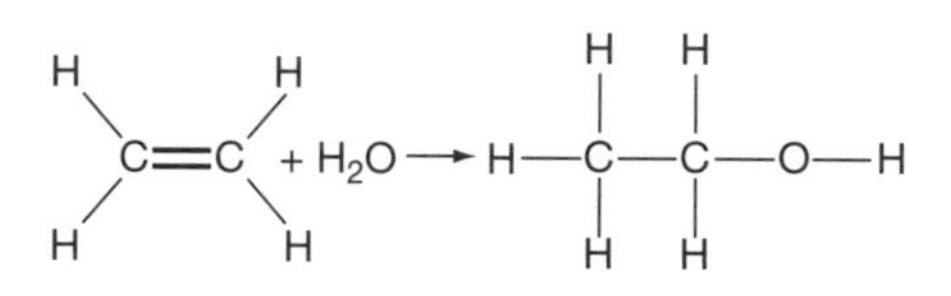

← Water (steam) adds to ethene to make ethanol

Ethanol by fermentation

Revised

- **Ethanol** can also be produced by **fermentation** with **yeast**, using **renewable resources**.
- This can be represented by:

sugar → carbon dioxide + ethanol

$$C_6H_{12}O_6 \rightarrow 2CO_2 + 2C_2H_5OH$$

- The process starts with a starchy material such as barley or, in hot countries like Brazil, sugar cane.
- The starch is extracted from the plants using hot water.
- Starch is a complex carbohydrate and it has to be broken down into sugars using enzymes.
- The sugar is then **fermented** at a warm temperature of about 35°C for several days.
- When the ethanol concentration reaches about 15% the **yeast dies** and fermentation stops.
- The ethanol has to be **concentrated** and **separated** from the mixture by **fractional distillation**.

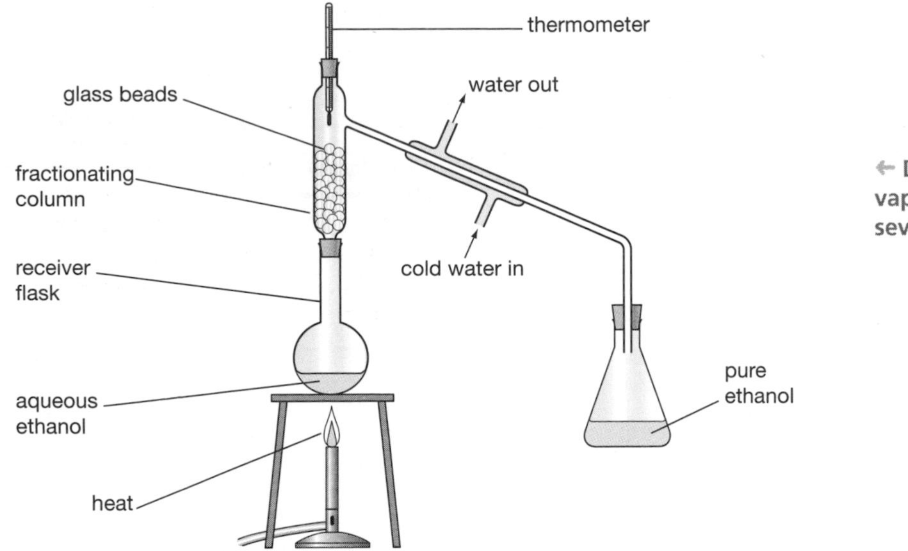

During fractional distillation the vapours condense and evaporate several times in the vertical column

Which process is best?

Revised

Using ethanol in car engines may produce less pollution than using petrol (see page 17). But is it better for us (in terms of the environment or economic cost) to use ethene (from crude oil) or to ferment plant material to make ethanol for use as a fuel?

Look at the table below, which shows **some** of the advantages and disadvantages of each process.

	Advantages	**Disadvantages**
Ethanol from ethene	95% conversion One-stage process	Ethene comes from crude oil — a non-renewable resource High energy cost of high temperature and pressure
Ethanol from fermentation	Cheap raw material Renewable plant material — conserves fossil fuels Low-cost process, because of low temperature	15% mixture must be concentrated and purified by distillation — extra cost Two-stage process

You could be asked to describe similarities and differences in:

- the economic costs
- the speed of production
- the impact on the environment
- the raw materials
- the purity of the final products
- the concentration of the ethanol produced
- the uses of each 'type' of ethanol
- whether the process can be made continuous or not, or if you have to wait for a batch to finish

Check your understanding

Tested

43 Explain why you can separate the ethanol from the fermentation mixture. *(2 marks)*

44 How is ethanol purified? *(1 mark)*

45 What processes are used to produce ethene from crude oil? *(1 mark)*

46 If you had a choice at a filling station between a fuel produced only from crude oil and one produced from plant material, which one would you choose? Justify your answer. *(4 marks)*

Answers online **Test yourself online** Online

Vegetable oils

Vegetable oils in food

Revised

Some fruits, seeds and nuts are rich in **oils** that can be **extracted**. This **extraction** is done by crushing the plant material and then removing the oil by **pressing** or in some cases by **distillation**. Then water and other impurities are also removed.

examiner tip

Remember: extraction of oils from fruit, seeds or nuts is done by **crushing and then pressing or distillation**.

Some common plant oils and the sources of the oil

Plant oil	Where oil is stored in the plant
Olive oil	Fruit
Rape oil	Seeds
Peanut oil	Nut (food store for seeds)
Avocado oil	Fruit
Sunflower oil	Seed
Palm oil	Fruit

- Vegetable oils are important **foods** because they provide **energy** and **nutrients**.
- Vegetable oils have **higher** boiling points than water and so can be used to **cook** foods at **higher temperatures** than by boiling in water.
- This produces **quicker** cooking and **different flavours** but increases the **energy** that the food **releases** when it is eaten.

Vegetable oils for fuels: biodiesel

Revised

- Vegetable oils are also important **fuels — biodiesel** is an example.
- Old oil from the chip shop can be used to make fuel. But **most biodiesel** is made from **crops** grown for that purpose.
- Biodiesel is **renewable** and **conserves crude oil** reserves.
- But, as well as **benefits**, there are **drawbacks** and **risks** of using vegetable oils to produce fuels.

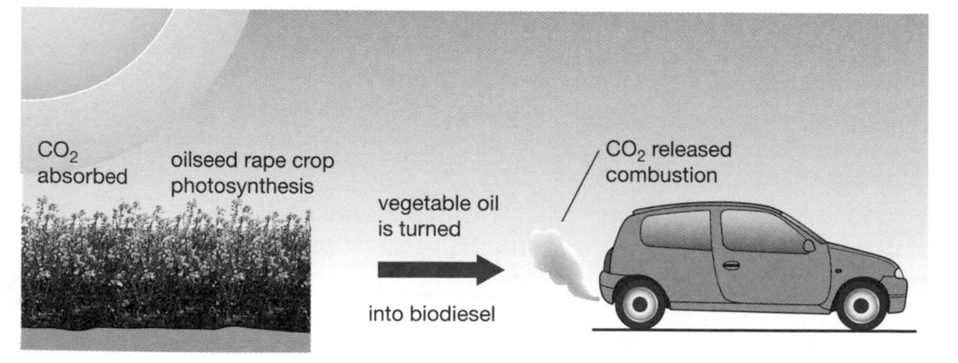

← **Burning biodiesel produces no CO_2 overall; CO_2 is absorbed by the plants from which the fuel is made**

Benefits	Drawbacks	Risks
No modification of diesel engine needed Can be mixed with ordinary diesel Much **less polluting** than ordinary diesel It **biodegrades** faster than ordinary diesel It is carbon neutral Biodiesel can be produced locally, benefiting the local economy	A lot of land is used to grow plants for fuel instead of food	Increased use of fertilisers Increased use of GM crops to withstand weedkillers Allergic reactions to chemicals used in the whole process

Emulsions

- **Oils** do not **dissolve** in water but they can be used to produce **emulsions**.
- **Emulsions** are special **mixtures** of oil and water that are thicker than oil or water.
- **Emulsions** have many uses that depend on their **special properties**.
- **Emulsions** provide better texture, coating ability and appearance than oil or water separately.
- **Salad dressings** (such as mayonnaise) and **ice creams** are emulsions.
- **Cosmetics** and **paints** are also emulsions — better **texture**, **coating ability** and **appearance** are important here, as well as in food.
- **Emulsifiers** stop the oil and water mixture separating out. In mayonnaise the emulsifier is egg yolk. French salad dressing (olive oil and vinegar) has no emulsifier and you can watch the two separate just after you have shaken them up.

MAYONNAISE

FRENCH DRESSING

→ **Emulsifiers keep emulsions emulsified**

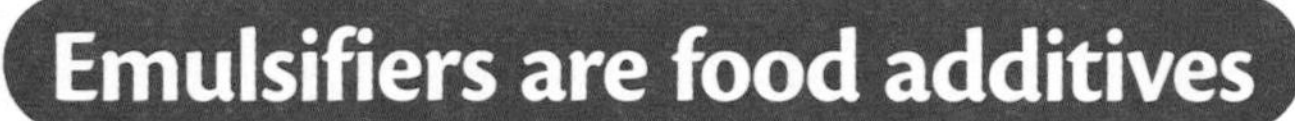

Emulsifiers are food additives

Revised

- Emulsifiers make emulsions **stick** to food, making it **taste** better and **look** more appetising than just oil and water.
- Additives permitted in the EU have E-numbers to show they have passed safety tests.
- But we have all heard of people who are allergic to certain **additives**.
- So while there are **benefits**, there are also **drawbacks** and **risks** involved in using **emulsifiers**.

How emulsifiers work

Revised

- Emulsifiers have **hydrophilic** and **hydrophobic** properties.
- An emulsifier molecule has a **hydrophilic** 'head' and a **hydrophobic** 'tail'.
- The head is strongly attracted to water — hence hydrophilic.
- The tail is strongly attracted to oil (it acts as if it is repelled by water — hence hydrophobic).

hydrophilic end

hydrophobic end

water-loving head

oil-loving tail

↑ **A typical emulsifier molecule has a 'head' that is attracted to water and 'tail' that is attracted to oil.**

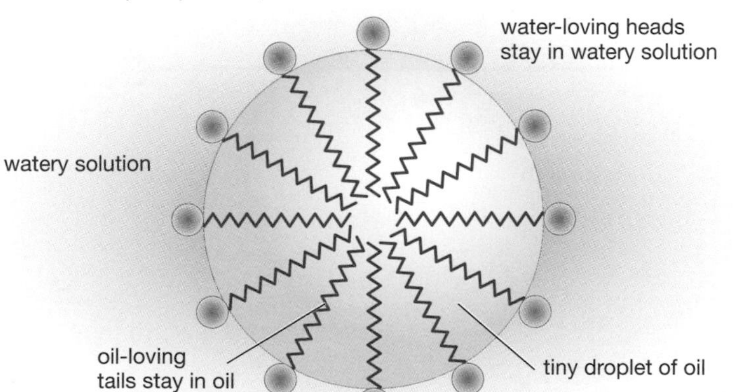

← **The emulsifier molecules penetrate tiny oil droplets and prevent them separating from the water**

examiner tip

'Hydrophilic' means 'attracted to water' and 'hydrophobic' is 'attracted to oil'. Avoid using the terms 'water loving' or 'water hating' in the exam.

Check your understanding

Tested

47 What is most biodiesel made from? *(1 mark)*

48 Biodiesel is carbon neutral. What does that mean? *(2 marks)*

49 What does an emulsifier do? *(1 mark)*

50 Describe an emulsifier molecule and explain how it keeps an emulsion stable. *(5 marks)*

Answers online **Test yourself online** Online

Saturated and unsaturated oils

- Vegetable oils that are **unsaturated** contain **carbon–carbon double bonds**. These can be detected by reacting with **bromine water**. (See page 19.)
- Fats and oils do not easily mix with bromine water, so it is useful to warm them in a water bath to help with the test.

An unsaturated fat has chains like this:

A saturated fat has chains like this:

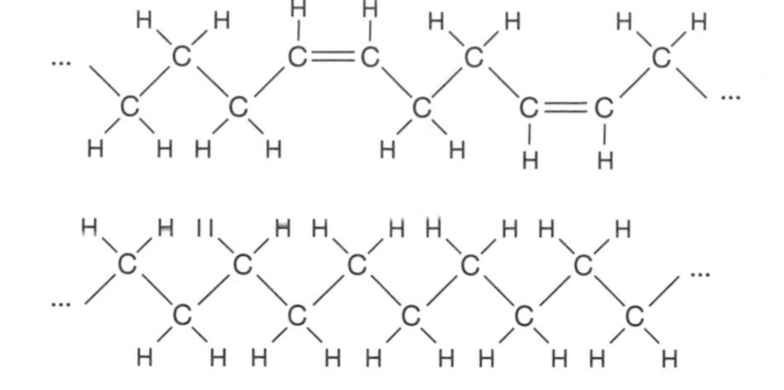

↑ **The structure of molecules of saturated and unsaturated fats**

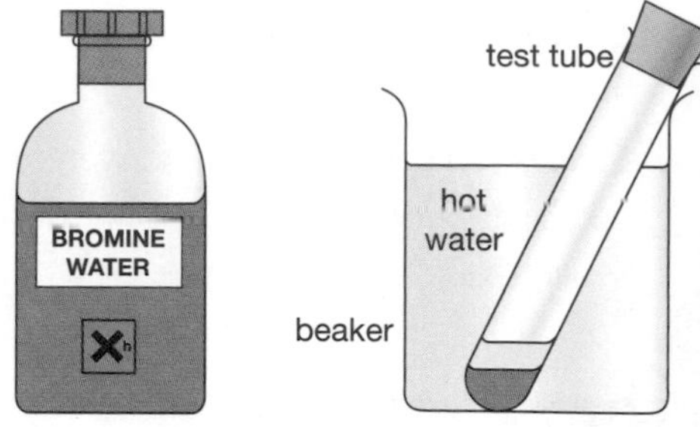

↑ **Testing for unsaturated oils or fats**

- **Vegetable oils** are **unsaturated** and are usually **liquid** at room temperature.
- They can be **hardened** by reacting them with **hydrogen** in the presence of a **nickel catalyst** at about **60°C**.
- **Hydrogen** adds to the double bonds and this is called **hydrogenation**.
- The **hydrogenated** oils have **higher melting points** so they are **solids** at room temperature, making them useful as spreads and in cakes and pastries.

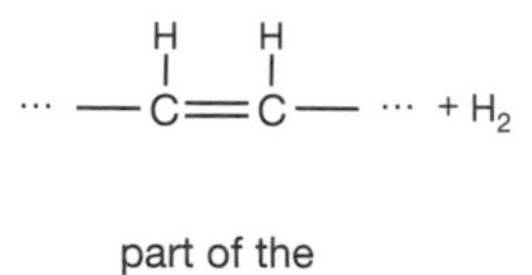

$$\cdots -\text{CH}=\text{CH}- \cdots + H_2 \rightarrow \cdots -\text{CH}_2-\text{CH}_2- \cdots$$

part of the oil molecule — part of the spread molecule

← **Hydrogenation of a plant oil**

Plant oils in margarine

Revised

The process of making margarine increases the amount of saturated fat in the product. Too much **saturated** fat in your diet can cause **health problems** — **saturated** fats can lead to heart disease and stroke.

Remember that:

- **too much** dietary **fat** and not enough exercise can lead to obesity
- **saturated fats** can lead to **heart disease** and **stroke**
- the ingredients label may also show other **additives** that have other effects on health (e.g. salt)

Some people say that margarine is still better for you than butter.

You can be asked to use the data in a table to **evaluate** the effects of using vegetable oils in foods and the impacts on diet and health.

Evaluate means you have to carefully read the data and deal with the **evidence**, using words from the question, for example '**use**', '**benefits**' and '**drawbacks**' — forget one and you will lose marks.

Comparing butter and different margarines

	Butter	Hard margarine	Soft margarine made with olive oil
Fat per 100 g (g)	81	81	59
Of which saturates (%)	64	20	24
Cholesterol per serving (mg)	30	0	0

examiner tip

Use a table to do your evaluations.

In this case calculate the mass of saturated and unsaturated fats in 100 g of each spread.

Calculation of fats in butter and different margarines

	Butter	Hard margarine	Soft margarine made with olive oil
Fat per 100 g (g)	81	81	59
Of which saturates (g)	52	16	14
Of which unsaturates (g)	29	65	45
Cholesterol per serving (mg)	30	0	0

Remember: **saturated** fats have **health risks** attached to them.

Evaluation can be done in the form of a table like this:

Evaluation of fats in butter and different margarines

	Use	Benefits	Drawbacks
Butter	Spread on bread; in cakes and pastries	Tastes nice	Unspreadable in cold weather; high fat content; higher in saturated fats than margarine
Hard margarine	In cakes and pastries	Low in saturates; no cholesterol	High fat content
Soft margarine made with olive oil	Spread on bread	Spreadable in all weathers; low fat; no cholesterol	—

Check your understanding

Tested

51 What is an unsaturated oil? *(1 mark)*

52 a) Describe how to test butter and a range of oils to see which is unsaturated. *(2 marks)*

b) What will you see happen if the fat or oil is unsaturated? *(2 marks)*

53 Explain why unsaturated fats are considered to be healthier. *(2 marks)*

54 Unsaturated fat can be turned into a form useful for baking.

a) What is the reaction called? *(1 mark)*

b) What is the other reactant? *(1 mark)*

c) What are the conditions needed for the reaction to happen? *(2 marks)*

Answers online **Test yourself online** Online

The Earth's crust

The Earth and its atmosphere provide **everything** we need: the Earth's crust, the atmosphere and the oceans are the **only source** of **minerals** and **other resources**.

The structure of the Earth

Revised

The Earth has a **layered structure**: a **core**, **mantle** and **crust** and is surrounded by the **atmosphere**.

- The **core** is very dense.
- The **mantle** is a thick band of mostly solid rock, but it can move very slowly.
- The **crust** is much thinner than the mantle and core.
- The **atmosphere** is very thin.

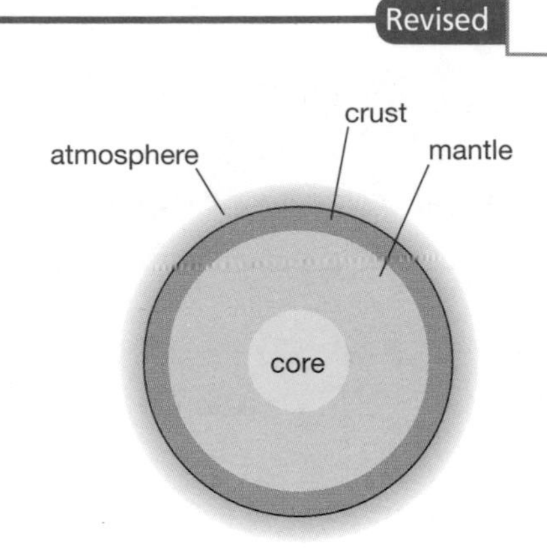

A section through the Earth

Scientists once thought that:

- as the Earth had **cooled down** it contracted
- the **shrinking** of the crust caused wrinkling on the surface
- this wrinkling is seen as **mountains**

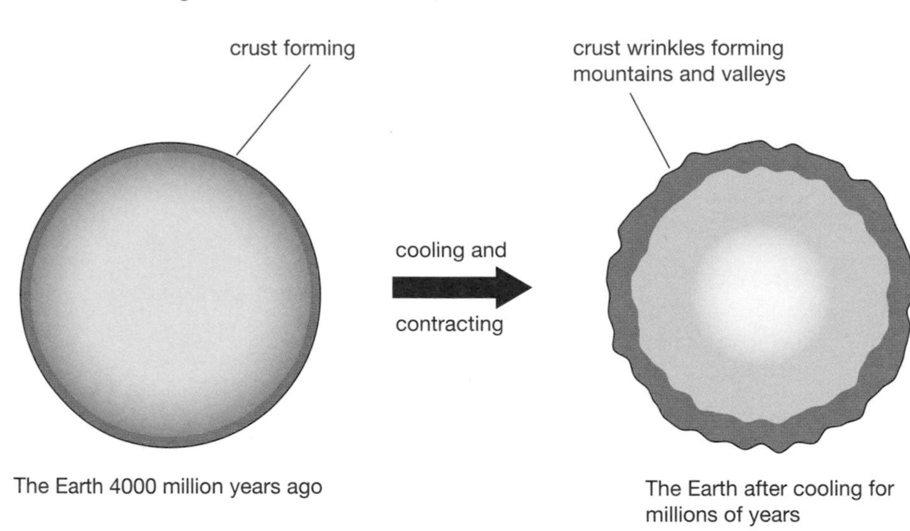

Early ideas about the formation of mountains on the Earth's surface

Then a meteorologist named Alfred Wegener came up with a radically different idea.

Continental drift

Revised

Wegener observed that on opposite sides of the Atlantic Ocean there were:

- **similar ancient rocks** and mountain chains
- very **similar fossils**
- a very **good fit** of the shape of the coastlines

He said this could be explained if:

- all the continents were once joined together as one **supercontinent**
- the supercontinent broke up and the continents had **slowly moved apart**

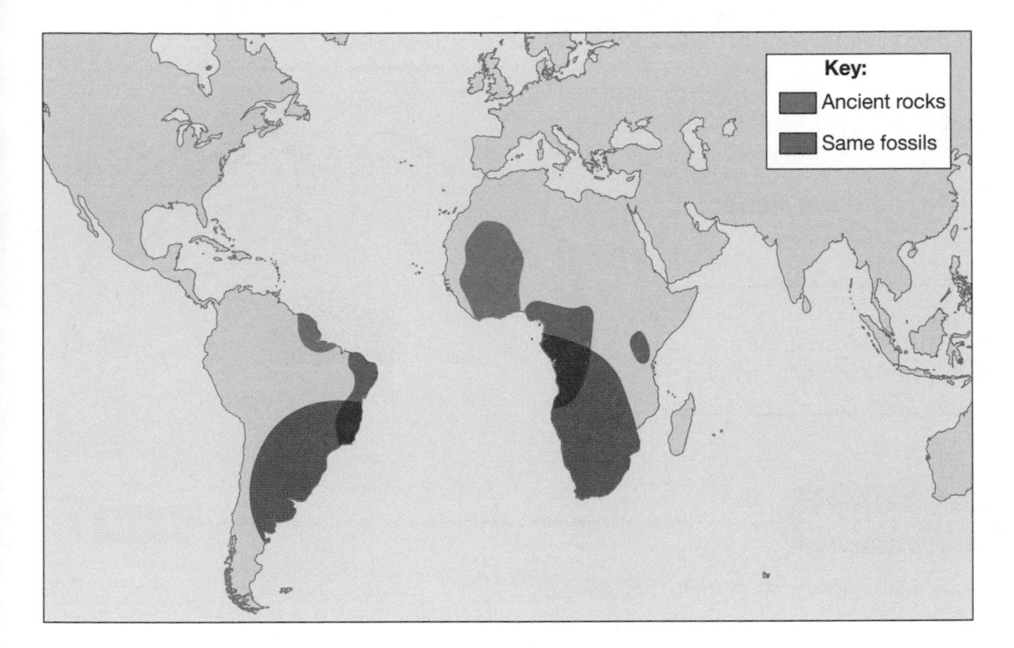

The 'jigsaw fit' and similar rock types and fossils of South America and Africa were used as evidence for Wegener's theory of continental drift

But Wegener's theory of crustal movement (**continental drift**) was not generally accepted for many years after it was proposed.

This was because:

- other scientists were **hostile** — after all, it had been proposed by a 'weatherman', not a geologist
- the geological experts had **other interpretations** of some of the evidence — a bridge of land between the continents could have allowed animals to pass freely
- Wegener **could not explain** how the continents could move through the solid ocean floor
- **nobody knew** what was at the **bottom of the oceans** between the continents

Today, scientists accept Wegener's theory because of **new evidence** about the **ocean floor**, new evidence about the **interior** of the Earth, and a theory to explain how the crust can move — **plate tectonics**). This is an example of how **scientific ideas** can **change** with time as scientists check each other's ideas and **evidence**.

Plate tectonics

Revised

The theory that replaced continental drift in the 1970s is called the **theory of plate tectonics**. This theory says:

- The Earth's **crust** and the upper part of the **mantle** are cracked into a number of large pieces called **tectonic plates**.

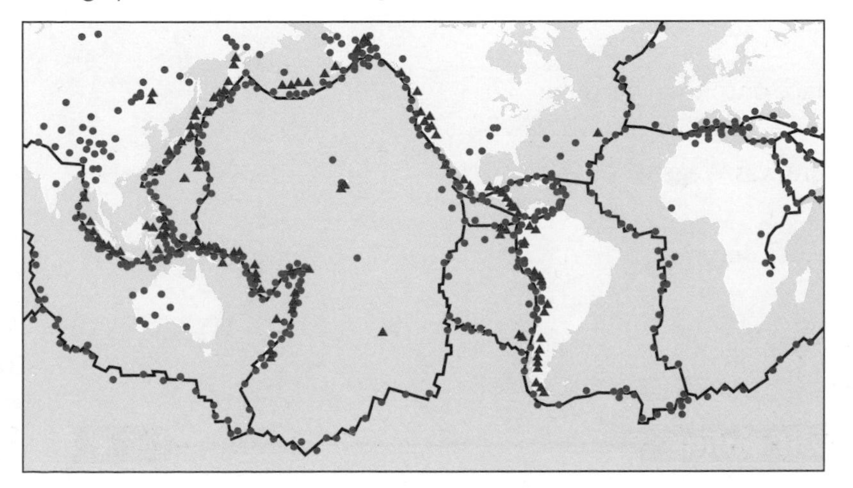

Key
plate boundaries
earthquakes
volcanoes

The tectonic plates and the pattern of volcanoes and earthquakes

- **Natural radioactive** processes deep inside the Earth produce heat.
- This heat causes **convection currents** within the mantle, even though it is mostly solid.
- The tectonic plates **move** because of the **convection currents**.

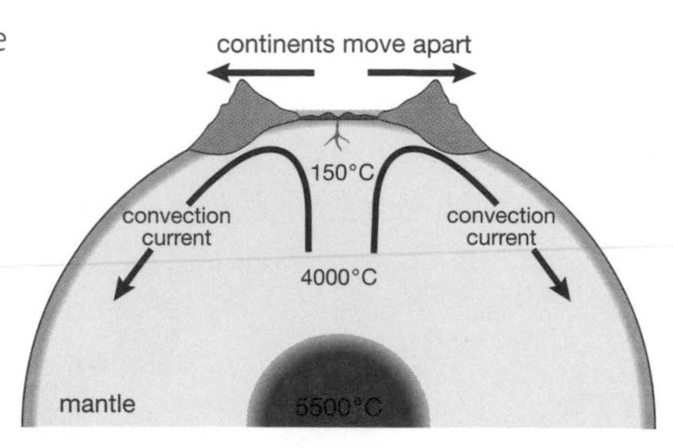

↑ Convection currents make tectonic plates move

examiner tip

Remember that the convection currents in the mantle are caused by **radioactive processes**, not nuclear reactions.

Earthquakes and volcanoes

Revised

- Usually the **tectonic plates** move really slowly — at **relative speeds** of a few **centimetres** per year.
- The movement causes stresses and strains to build up between the plates.
- **Earthquakes** and/or **volcanic eruptions** occur at the boundaries between **tectonic plates** as the result of **sudden** and **disastrous** movements of the plates.

Plate tectonic theory predicts **where** earthquakes and volcanic eruptions are likely to occur. But scientists cannot accurately predict **when** stresses and strains will be released.

They can measure:

- the forces in rocks
- tiny movements and bulges in the Earth's crust
- shock waves from small earthquakes that happen before the main one
- movement of the tectonic plates using global positioning systems

These may give a **warning sign** that an earthquake or eruption is **likely**, but not a definite prediction of when.

Check your understanding

Tested

55 Draw a labelled diagram clearly showing the structure of the Earth. *(4 marks)*

56 Describe how a new scientific theory becomes accepted by other scientists. *(2 marks)*

57 How were mountains on the Earth's surface once thought to be formed? *(1 mark)*

58 Which part of the Earth's surface structure was Wegener's theory unable to explain? *(1 mark)*

59 In which part of the Earth do tectonic plates occur? *(1 mark)*

60 How fast do tectonic plates move? *(1 mark)*

61 Explain the movement of tectonic plates. *(4 marks)*

62 Where and why do earthquakes occur? *(1 mark)*

Answers online **Test yourself online**

Online

The Earth's atmosphere

Mainly nitrogen and oxygen

Revised

For 200 million years, the proportions of different gases in the **atmosphere** have been much the same as they are today:

- about four-fifths (80%) **nitrogen**
- about one-fifth (20%) **oxygen**
- small proportions of various other gases, including **carbon dioxide**, **water vapour** and **noble gases**

→ The gases in our atmosphere

Where did they come from?

Revised

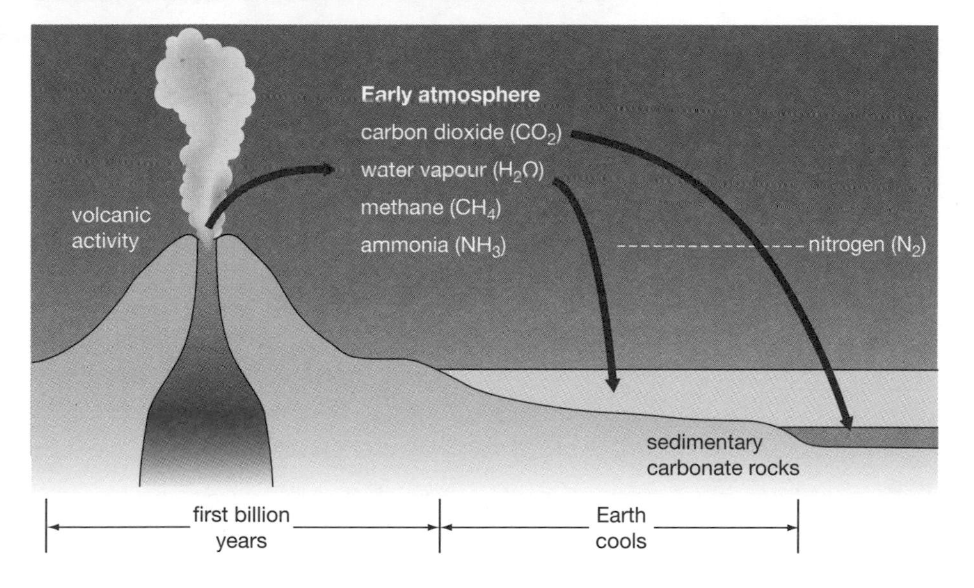

← Volcanoes released the gases in the Earth's early atmosphere

During the first billion years of the Earth's existence, there was intense **volcanic activity**. These volcanoes released the gases that formed the early atmosphere:

- carbon dioxide (CO_2) — main component of the early atmosphere
- methane (CH_4) — small proportion
- ammonia (NH_3) — small proportion
- water vapour (H_2O)

As the Earth cooled, the **water vapour condensed** to form the oceans.

There are several theories about how the atmosphere was formed.

One theory suggests that during this early period the Earth's atmosphere was **mainly carbon dioxide** and there would have been **little or no oxygen** gas. This would have been like the atmospheres of **Mars** and **Venus** today.

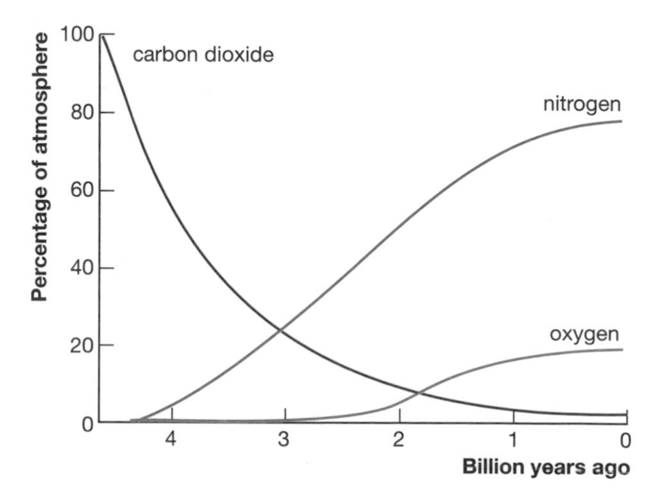

↑ How the composition of the Earth's atmosphere has changed

The formation of life

Revised

- The early atmosphere contained all the right elements to generate life.
- There are many theories as to how life was formed **billions** of years ago.

One involves the interaction between hydrocarbons, ammonia and lightning. In the famous Miller–Urey experiment the conditions of the early atmosphere were reproduced in the laboratory and a '**primordial soup**' of amino acids was formed.

However, there is no specific theory about how **organisms** could develop from such a mixture. Neither is there any **fossil evidence** of the process — only soft tissues would have been formed.

This is not the only theory but we still **do not know** how life was formed.

How carbon dioxide and oxygen levels changed

Revised

Plants and algae produced the oxygen that is now in the atmosphere.

Plants and **algae** use **photosynthesis** to:

- **consume carbon dioxide** — atmospheric carbon dioxide levels decrease
- **produce oxygen** — atmospheric oxygen levels increase

examiner tip

Photosynthesis in plants increases atmospheric oxygen.

Photosynthesis in plants decreases atmospheric carbon dioxide.

These statements score 2 marks **each** — 'photosynthesis' scores twice.

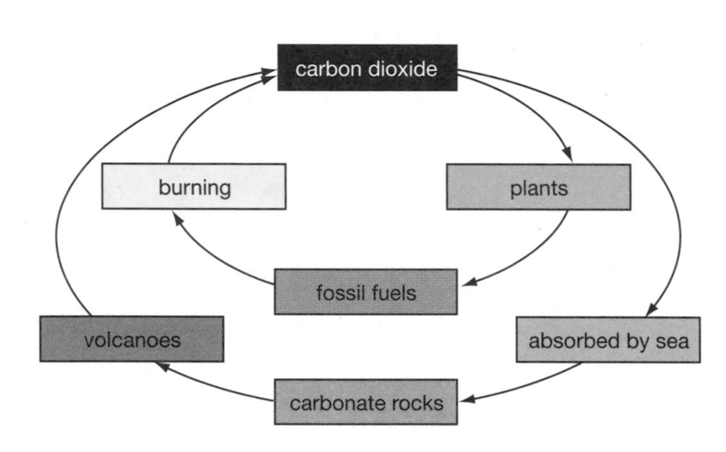

↑ Carbon dioxide moves into and out of the atmosphere

- Most of the carbon from the carbon dioxide in the air gradually became locked up in sedimentary rocks as **carbonates** and **fossil fuels**.
- When carbon dioxide dissolves in the oceans it can be used to form limestone and the shells and skeletons of marine organisms.
- The remains of marine plants and animals form fossil fuels that contain carbon and hydrocarbons.
- Nowadays the release of carbon dioxide by **burning fossil fuels** increases the level of carbon dioxide in the atmosphere. This **increase** in **carbon dioxide** is thought to be causing **global warming**.
- The oceans also act as a **reservoir** for **carbon dioxide**, but increased amounts of carbon dioxide absorbed by the oceans have an **impact** on the **marine environment**.

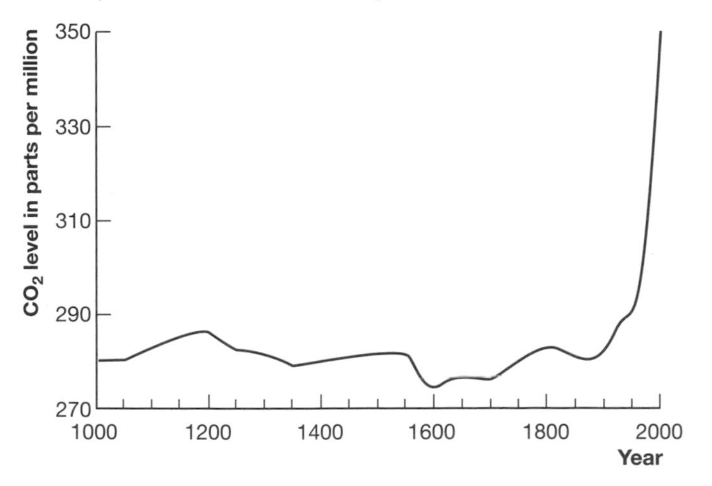

← The amount of carbon dioxide in the atmosphere has increased over the past 1000 years

Fractional distillation of air

Revised

- Air is a mixture of gases with **different** boiling points.
- It can be **fractionally distilled** to provide a source of raw materials that are used in a variety of industrial processes.
- The water vapour and carbon dioxide are removed before the remaining mixture is liquefied by cooling.
- The liquid mixture is allowed to warm up.
- The gas with the lowest boiling point leaves the liquid first.

Uses of gases in the air — you do not need to remember the boiling points

Gas	Boiling point (°C)	Use
Nitrogen	−196	Inert gas for packaging
Oxygen	−183	Rocket propellant, welding, medical use
Argon	−186	Filling light bulbs, inert gas for arc welding
Carbon dioxide	Sublimes at −78	Dry ice refrigerant
Helium	−269	Balloon gas, inert gas for welding

examiner tip

If you are asked to 'explain' something you should give a fact **and** then give a consequence of that fact. For example: Plants produce oxygen, **therefore** oxygen levels in the atmosphere rise.

Check your understanding

Tested

63 What are the proportions of nitrogen and oxygen in the atmosphere today? *(2 marks)*

64 Describe how the proportions of oxygen and carbon dioxide in the Earth's atmosphere have changed over the past 4 billion years. *(4 marks)*

65 Use the data in the table to calculate the percentage change of carbon dioxide in the atmosphere since 1750. *(1 mark)*

Atmospheric CO_2 since 1750

Year	1750	1800	1850	1900	1950	2000
Concentration of carbon dioxide in the atmosphere (% by volume)	0.0278	0.0282	0.0288	0.0297	0.0310	0.0368

66 The early atmosphere was like the current atmosphere of which two planets? *(1 mark)*

67 Why do we not know how life was formed? *(2 marks)*

68 What environmental impact may be caused by increasing carbon dioxide levels in the atmosphere? *(1 mark)*

69 Explain why argon is used in filament lamps. *(2 marks)*

Answers online **Test yourself online**

Online

Structure and bonding (1)

- **Compounds** are substances in which atoms of two or more elements are chemically combined.
- **Chemical bonding** involves either **transferring** or **sharing** electrons in the **outer** shells (highest occupied energy levels) of atoms in order to achieve the electronic structure of a noble gas.

Ionic bonding

Revised

When atoms form chemical bonds by **transferring electrons**, they form **ions**.

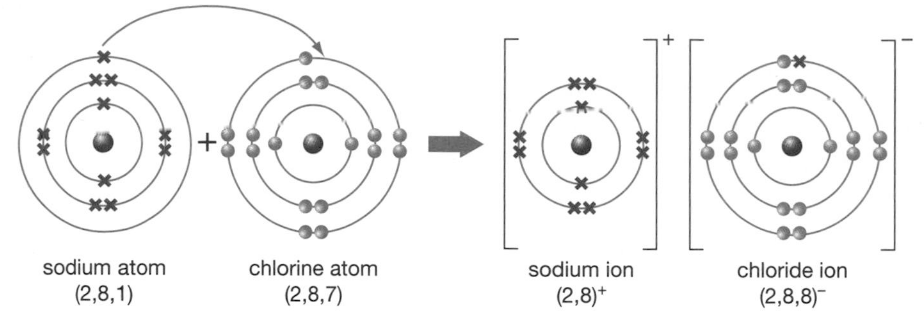

Ionic bonding and electron transfer

- Atoms that lose electrons become positively charged ions — more protons than electrons.
- Atoms that gain electrons become negatively charged ions — fewer protons than electrons.

Look at the table. Atoms gain or lose electrons to form ions having **the electronic structure of a noble gas** (group 0).

examiner tip

Pair up the electrons in your diagrams — it helps make sure that you have the right number in the right shells.

Group number determines the charge on simple ions

Group	1	2	3	4	5	6	7	0
Elements in period 3	Na	Mg	Al	Si	P	S	Cl	Ar
Electron structure	2,8,1	2,8,2	2,8,3	2,8,4	2,8,5	2,8,6	2,8,7	2,8,8
Number of electrons in outer shell	1	2	3	4	5	6	7	8
Common ion	Na^+	Mg^{2+}	Al^{3+}	No ion	P^{3-}	S^{2-}	Cl^-	No ion
Electron structure of ion	2,8	2,8	2,8	–	2,8,8	2,8,8	2,8,8	–

We draw the same type of diagram to represent the electronic **structure** of ions as we do for atoms but we must include **brackets** and the **charge of the ion**. We also use brackets and the charge when writing the electronic **configuration** of ions, for example $(2,8)^+$.

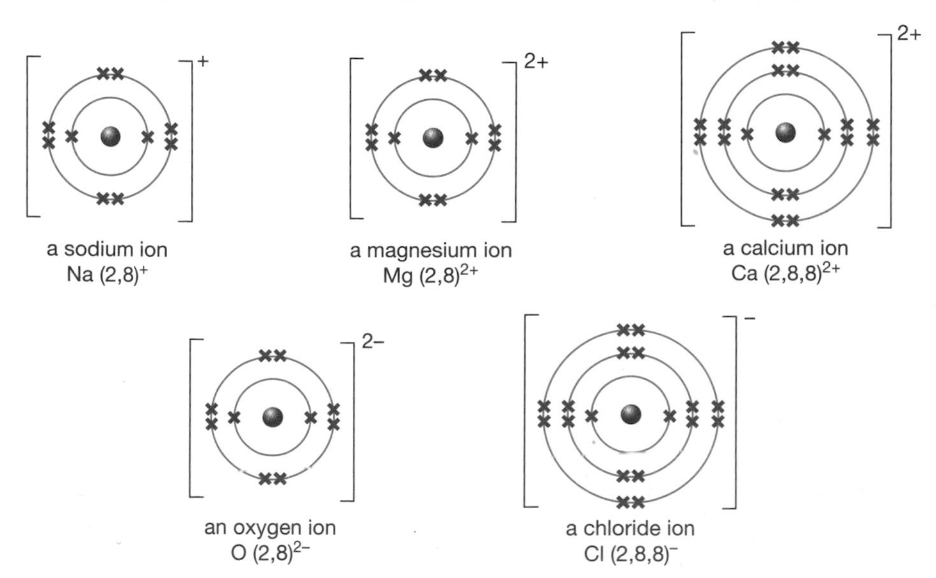

Ionic reactions

Revised

Group 1 elements, the alkali metals, all react with non-metal elements to form ionic compounds in which the **metal ion** has a **single positive charge**, for example a sodium ion, Na^+.

Group 7 elements, the halogens, react with the alkali metals to form ionic compounds in which the **halide ions** have a single negative charge, for example a chloride ion, Cl^-.

The typical ionic reaction between sodium metal and chlorine gas can be represented in a number of ways:

- a word equation: sodium + chlorine → sodium chloride
- a balanced symbol equation: $2Na(s) + Cl_2(g) \rightarrow 2NaCl(s)$
- as in the diagram at the start of this section, showing ionic bonding is electron transfer
- as a 'dot-and-cross' diagram like this:

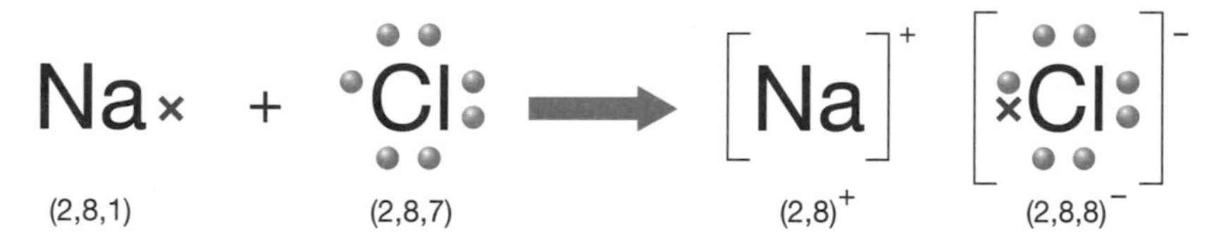

← **A 'dot-and-cross' diagram for the reaction between sodium and chlorine**

The reactions between the elements in groups 1 and 7 are all very exothermic. Some are explosive in nature and they all produce **soluble white solids**.

Generally the reactions can be represented like this:

$2M + X_2 \rightarrow 2MX$ (M represents the metal and X represents the halogen)

Ionic compounds

Revised

- An ionic compound is a **giant structure** of ions held together by **strong forces of attraction** between **oppositely charged ions**.
- These forces act in **all directions** in the lattice: this is called **ionic bonding**.
- In the formula for any ionic compound, the sum of the charges on the ions **must** be zero.
- Check this by looking at KCl (K^+Cl^-), Na_2O ($Na^+{}_2O^{2-}$) and $CaBr_2$ ($Ca^{2+}Br^-{}_2$).
- See page 42 for more about ionic structure and bonding.

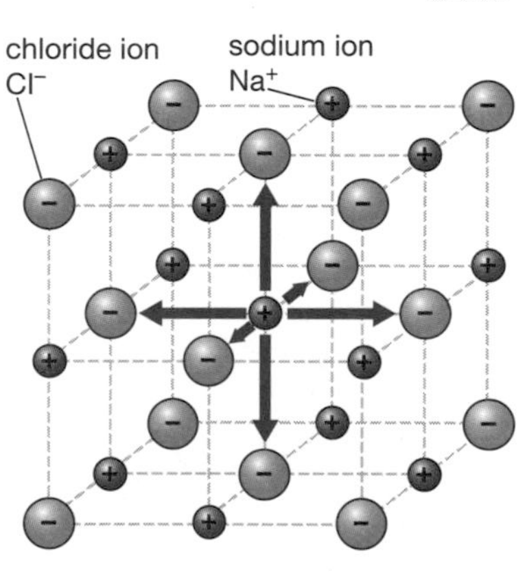

↑ **The ionic bond acts in all directions**

examiner tip

To impress any examiner just write this in the correct place on your paper:
'An ionic bond is an omnidirectional electrostatic **force of attraction** between **oppositely charged ions**.'

Check your understanding

Tested

1 Write the formulae of the ionic compounds formed between the following pairs of ions:
  a) Na^+ and O^{2-} *(1 mark)*
  b) Mg^{2+} and O^{2-} *(1 mark)*
  c) Ca^{2+} and Cl^- *(1 mark)*

2 Draw diagrams to show clearly the transfer of electrons in the formation of:
  a) calcium oxide, CaO *(2 marks)*
  b) magnesium chloride, $MgCl_2$ *(2 marks)*

3 Represent the reaction between lithium and fluorine in four ways:
  a) a word equation *(1 mark)*
  b) a balanced symbol equation *(2 marks)*
  c) an electron transfer diagram *(2 marks)*
  d) a dot-and-cross diagram *(2 marks)*

Answers online **Test yourself online** Online

Structure and bonding (2)

- **Chemical bonding** involves either **transferring** or **sharing** electrons in the **outer** shells (highest occupied energy levels) of atoms in order to achieve the electronic structure of a noble gas.

Covalent bonding

Revised

When atoms **share** pairs of **electrons**, they form **covalent** bonds. These **bonds between atoms** are **strong**.

Some covalently bonded substances consist of **simple molecules** such as

- hydrogen, H_2
- chlorine, Cl_2
- oxygen, O_2
- hydrogen chloride, HCl (this dissolves in water to form hydrochloric acid)
- water, H_2O
- ammonia, NH_3
- methane, CH_4

The **molecules** are discrete, or **separate** from each other.

Others have **giant covalent structures** (macromolecules), for example diamond and silicon dioxide. (See page 40 for more on macromolecules.)

Name and formula	Structural formula	Model of structure
Hydrogen, H_2	H—H	
Oxygen, O_2	O=O	
Water, H_2O	H—O—H	
Methane, CH_4	H—C—H with H above and H below C	
Hydrogen chloride, HCl	H—Cl	
Chlorine, Cl_2	Cl—Cl	
Carbon dioxide, CO_2	O=C=O	
Iodine, I_2	I—I	

↑ The formulae and structure of some simple molecular substances

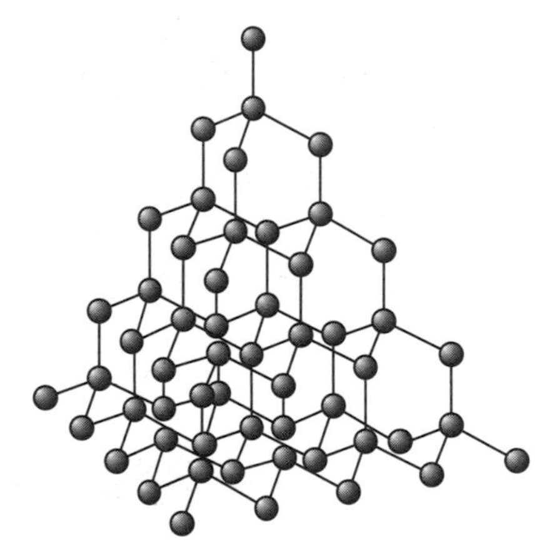

↑ Fragment of the diamond giant structure. Each carbon atom is linked to four others

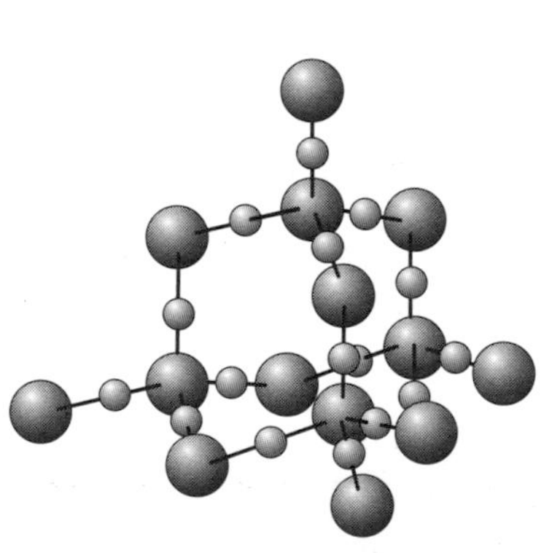

↑ Fragment of the silicon dioxide giant structure. The silicon atoms are arranged the same as the carbon atoms in diamond, but there is one oxygen atom between them

Dot-and-cross diagrams

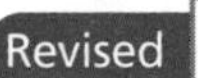

Dot-and-cross diagrams are used to represent covalent bonds in molecules. You show only the **outer electrons** of the atoms involved, but you can include the electron shell as a circle if you like.

It helps to show the electrons of the different atoms as either dots or crosses.

A pair of electrons (a covalent bond) is also represented by a single line between two atoms: H–H. A double bond (two pairs of electrons) as in oxygen, is shown like this: O=O.

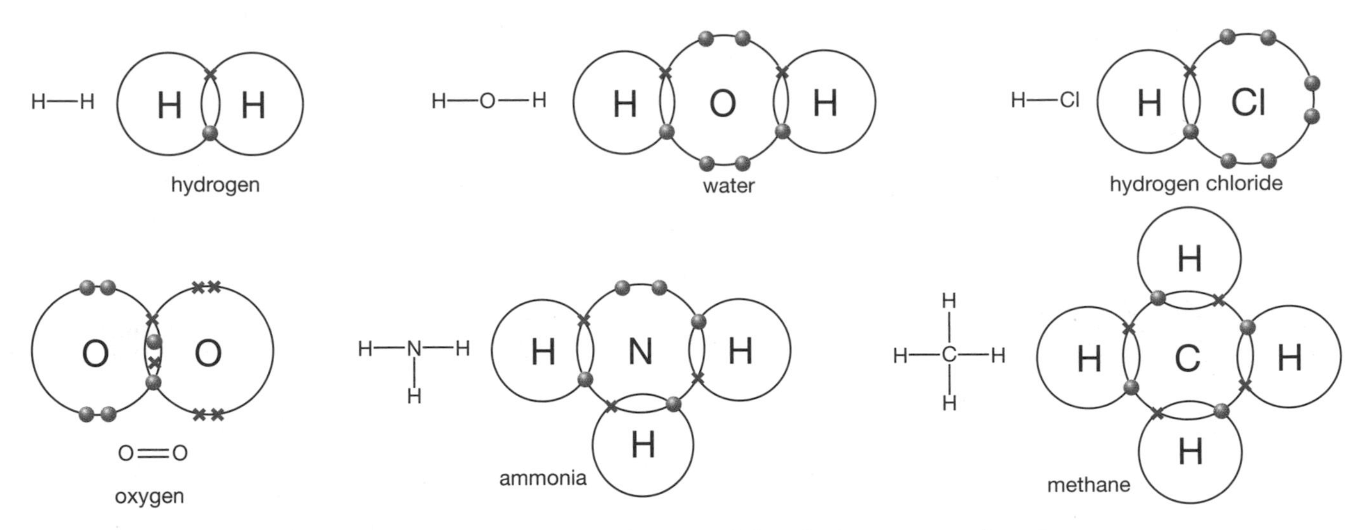

↑ Covalent bonding in some simple molecules. Notice that in each bond, electrons are shared (one from each atom)

examiner tip

In dot-and-cross diagrams each atom, except hydrogen, is surrounded by four electron pairs. But remember that the oxygen molecule has a double bond, so two of the pairs are shared.

Metallic bonding

Revised

Metals consist of **giant structures** of atoms arranged in a **regular pattern**.

The electrons in the highest occupied energy levels (outer shell) of metal atoms are **delocalised** and **free to move** through the whole structure. This corresponds to a structure of **positive ions** with **electrons** between the ions holding them together by **strong electrostatic attractions**.

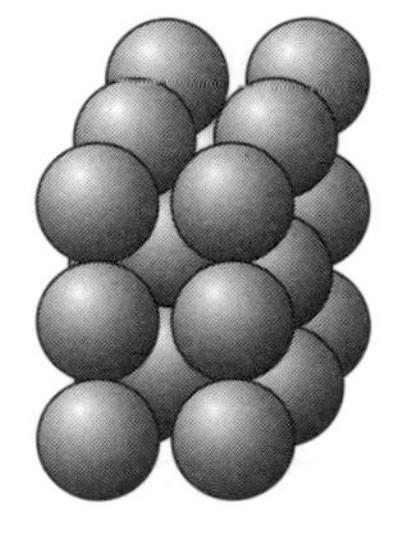

↑ Metals have giant structures

examiner tip

The delocalised electrons in metals move through the lattice.

Both levels: learn 'metals consist of giant structures of atoms'.

For higher level: 'the electrons are delocalised so that, effectively, positive ions are formed'.

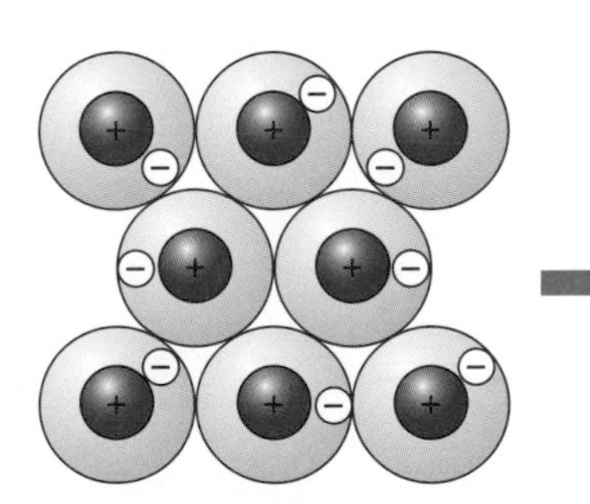

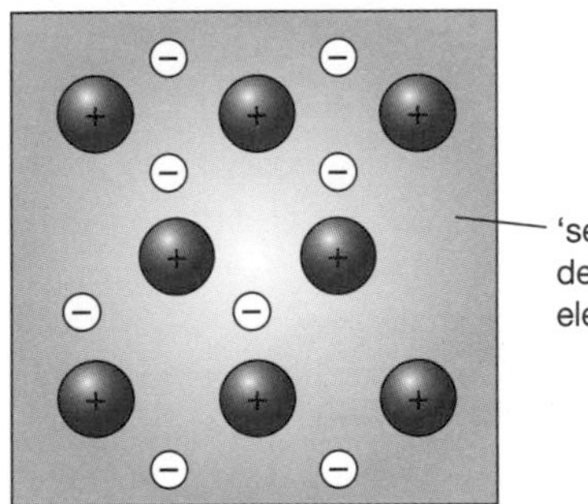

← Metallic bonding: attraction between negative delocalised electrons and positive ions result in strong forces of attraction

Check your understanding

Tested

4 Give the names and formulae of three materials that have simple molecular structures. *(1 mark per name, 1 mark per formula)*

5 Draw dot-and-cross diagrams to show clearly the formation of covalent bonds in:

a) phosphorus trichloride, PCl_3 *(2 marks)*

b) chlorine oxide, Cl_2O *(2 marks)*

6 Describe the structure of metals. *(2 marks)*

7 How are the ions in metals held together? *(2 marks)*

Answers online **Test yourself online** Online

Molecules

Substances that have simple molecular, giant ionic and giant covalent structures have very different properties.

Ionic, covalent and metallic **bonds** are **strong**. However, the **forces between** molecules are **weaker**.

Simple molecules

Revised

- Substances that consist of **simple molecules** are **gases**, **liquids** and **solids** that have relatively low melting points and boiling points.
- Simple molecules have a defined number of atoms in them — they are not giant structures.
- Substances that consist of **simple molecules** have only **weak forces** between the molecules (intermolecular forces).
- It is these **intermolecular forces** that are **overcome**, not the covalent bonds, when a substance **melts or boils**. It is important to remember that the **covalent bonds** (between the atoms in the individual molecules) **remain intact** — they are not so easily broken.
- **Intermolecular forces** are **weak** in comparison with covalent bonds.

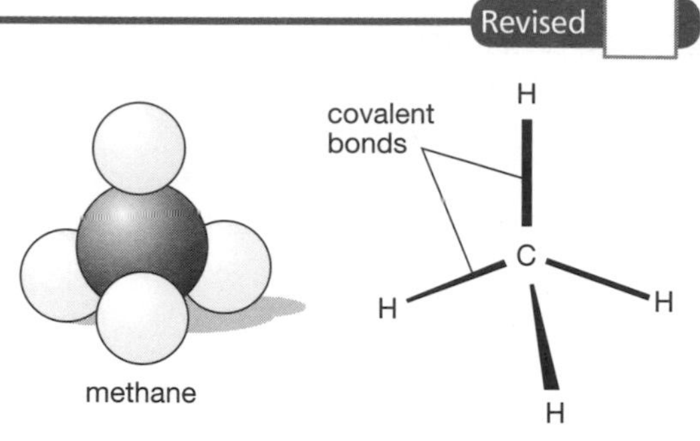

↑ **Methane (CH_4) is a simple molecular substance. In methane the atoms are held together by covalent bonds, which are strong**

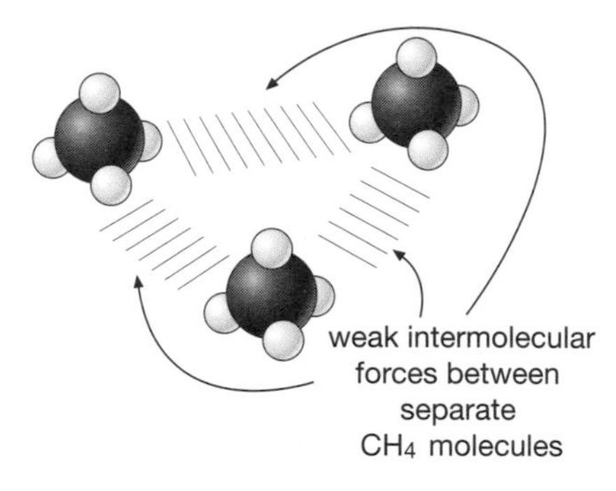

↑ **Intermolecular forces between molecules of methane are weak**

examiner tip

Remember: intermolecular forces are weak — they can be overcome easily. Covalent bonds are strong — they can be broken only with difficulty.

- Substances that consist of **simple molecules** do **not conduct electricity** because the molecules do not have an overall electric charge.

Covalent structures

Macromolecules (giant molecules)

Revised

- Atoms that **share electrons** can also form giant structures or **macromolecules**.
- **Diamond** and **graphite** (forms of carbon) and silicon dioxide (**silica**) are examples of **giant covalent** structures (lattices) of **atoms**.
- All the atoms in these structures are **linked** to **other** atoms by strong **covalent bonds** and so **macromolecular** substances have very **high melting points**.
- This is because melting (allowing the atoms to move freely) breaks all the (strong) covalent bonds that link them — this requires a lot of energy.

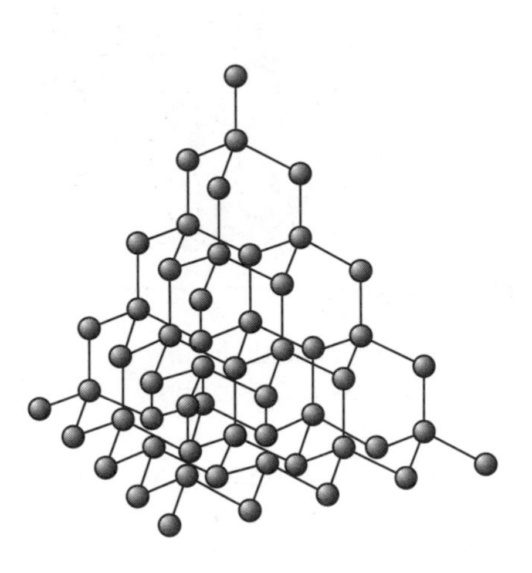

↑ The structure of diamond. Each grey ball represents a carbon atom and each line represents a (strong) covalent bond

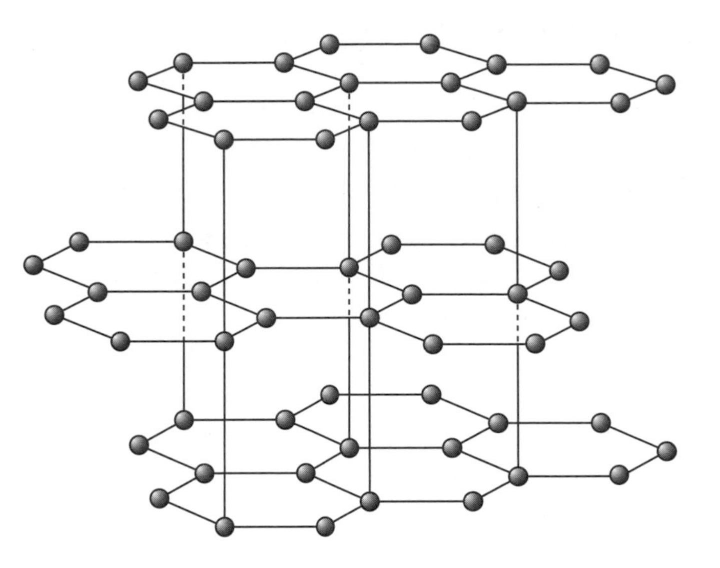

↑ The structure of graphite. The carbon atoms are arranged in layers. There are weak intermolecular forces between the layers

- In **diamond**, each carbon atom forms **four** covalent bonds with other carbon atoms in a **giant covalent structure**; this structure makes diamond very **hard**.
- In **graphite**, each carbon atom bonds to only **three** others, forming layers.
- In **graphite** the layers are free to slide over each other (like a pack of cards) because there are no covalent bonds between the layers and so graphite is **soft and slippery**.
- The layers in **graphite** are able to slide over each other because there are **weak intermolecular forces** between them.
- There is one electron from each carbon atom in **graphite** not used in bonding; it is **delocalised** (remember about delocalised electrons in metals — page 39). These delocalised electrons allow graphite to **conduct heat** and **electricity**.
- Carbon can also form **fullerenes** with different numbers of carbon atoms.
- Fullerenes can be used for drug delivery into the body, in lubricants, as catalysts, and in **nanotubes** for reinforcing materials (e.g. in tennis rackets).

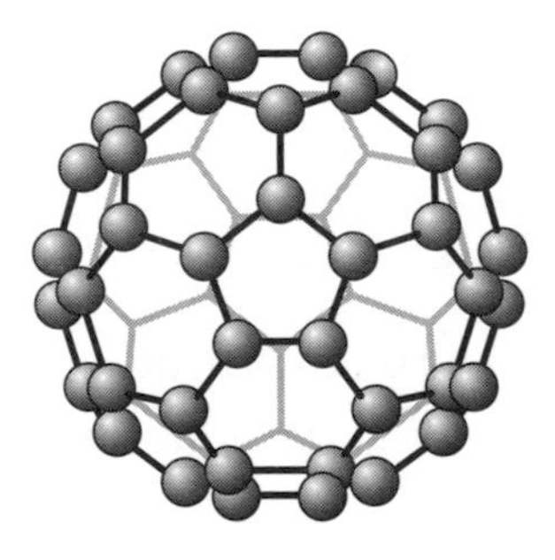

↑ The original C_{60} buckyball — a fullerene based on rings of carbon atoms

Check your understanding

Tested

8 Give two properties of simple molecular substances. *(2 marks)*

9 Explain why a kettle of water has water molecules in the gas immediately above the boiling surface and not hydrogen or oxygen molecules. *(4 marks)*

10 What types of particles form a giant covalent lattice? *(1 mark)*

11 Why is graphite sometimes used as a lubricant in hot machinery? *(2 marks)*

12 Why can diamond be used for saw and drill tips? *(1 mark)*

13 Diamond and graphite are both forms of diamond with giant covalent structures. Explain why diamond is very hard but graphite is a soft electrical conductor. *(3 marks)*

Answers online **Test yourself online**

Online

Ionic compounds

A typical example of an ionic compound is sodium chloride (NaCl), common salt.

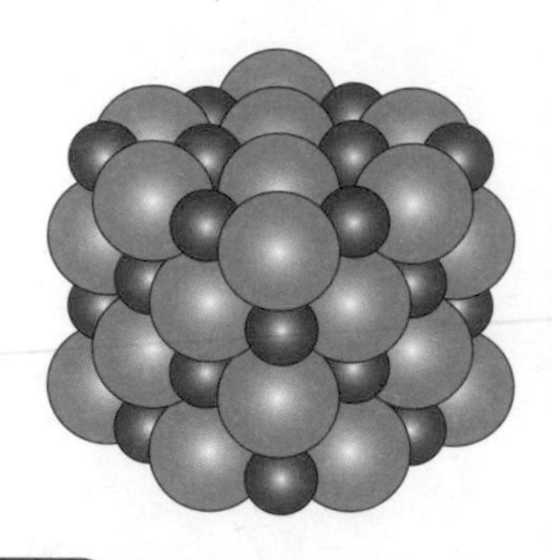

→ The structure of sodium chloride is typical of ionic compounds. The red balls represent sodium ions (Na^+), the green ones represent chloride ions (Cl^-)

- Ionic compounds have regular structures (**giant ionic lattices**) in which there are strong **electrostatic** forces in **all directions** between oppositely charged **ions**.
- These compounds have **high melting points** and **high boiling points**. This is because melting involves **breaking** the many (**strong**) ionic bonds. This requires large amounts of **energy**.
- When **melted**, or **dissolved** in water, ionic compounds conduct electricity because the **ions** are free to **move** and **carry** the **current**.

examiner tip

Melting produces a liquid; dissolving produces a solution (in ionic **and** covalent materials).

Ionic compounds have only ions in them; do **not** mention molecules or atoms.

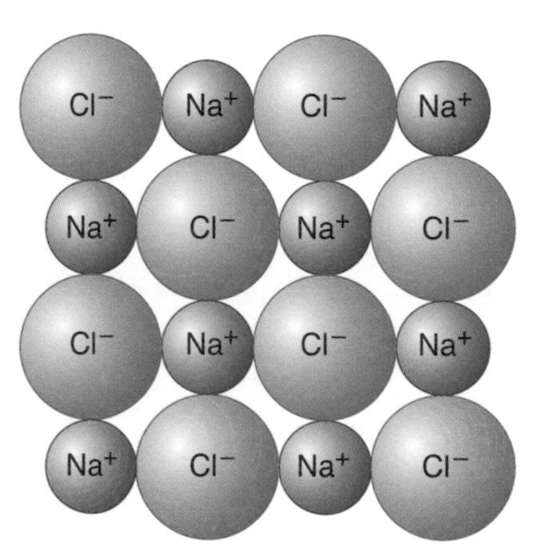

↑ Arrangement of ions in one layer of a sodium chloride (salt) crystal

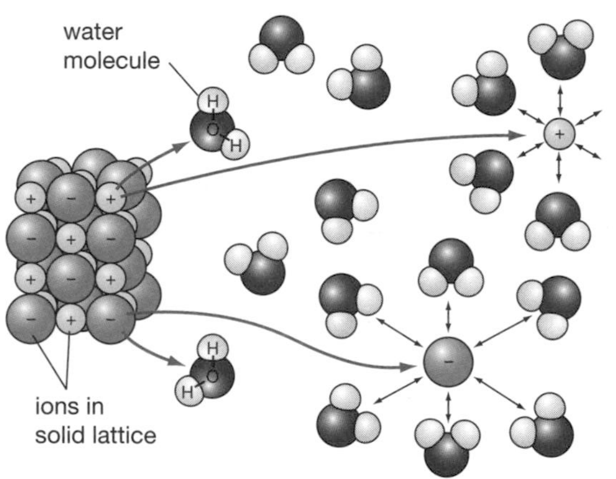

↑ Dissolved particles (in this case they are ions) are surrounded by solvent molecules (water) and are free to move

Metals

- Metals **conduct heat** and **electricity** because of the **delocalised electrons** in their structures.
- Delocalised electrons are **free** to move through the whole metal structure.
- The **conduction** of electricity in metals depends on this freedom of **movement** of electrons.
- The layers of atoms in metals are able to **slide** over each other, so metals can be **bent** and **shaped**.

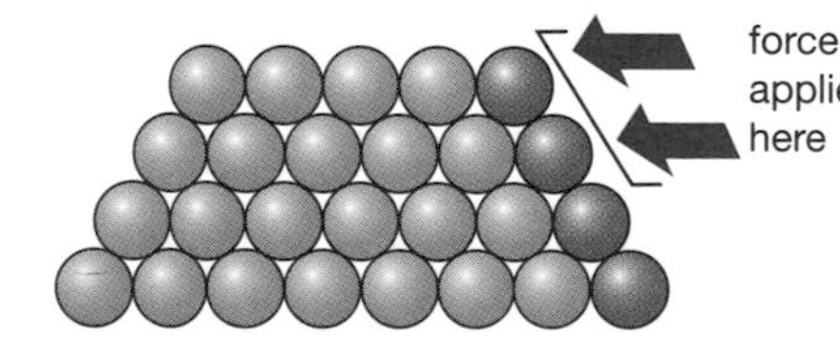

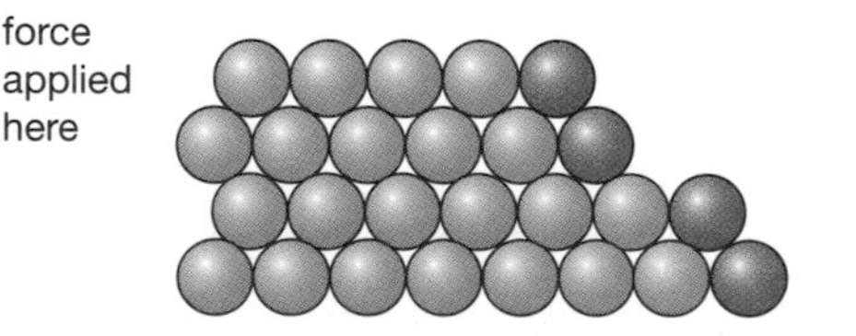

← Metals change shape when an applied force causes layers of atoms to slip over each other

Alloys

Revised

- Alloys are usually made from two or more different metals.
- The different-sized atoms **distort** the layers in the **structure**, making it more difficult for them to slide over each other and so making alloys **harder** than pure metals.

examiner tip

Alloys are harder than pure metals because their layer structures have been distorted.

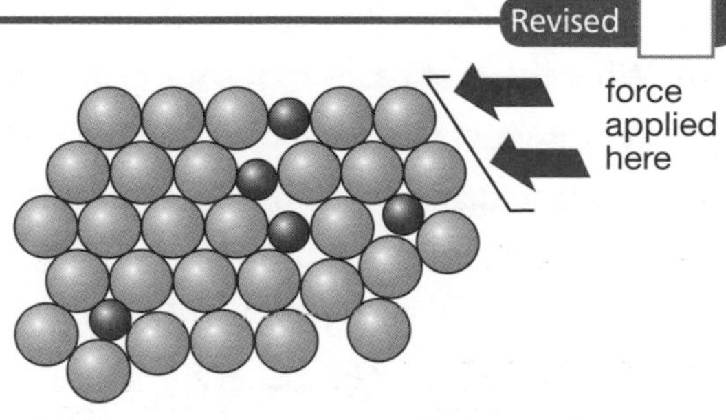

↑ Slip cannot occur so easily in an alloy, because the structure has been distorted

Shape memory alloys

Revised

- Alloys can be designed to have properties for specific uses by varying the amounts of other metals that are added.
- Shape memory alloys can **return** to their **original shape** after being deformed, for example Nitinol, which is used in dental braces and some spectacle frames.
- Nitinol is an alloy of nickel and titanium that has two different states — at different temperatures.
- With spectacles and dental braces the metal is normally at a temperature at which it always returns to its original shape, so there is no need to heat them.

Polymers

The properties of polymers depend on **what they are made from** and the **conditions** under which they are made. For example, low-density (LD) poly(ethene) and high-density (HD) poly(ethene) are produced using different **catalysts** and **reaction conditions**.

Thermosoftening and thermosetting polymers

Revised

- **Thermosoftening** polymers consist of **individual**, tangled polymer chains.
- **Thermosetting** polymers consist of polymer chains with **cross-links** between them so that they do not melt easily when they are heated.
- **Thermosoftening** polymers have low melting points because the weak intermolecular forces between the polymer chains can be overcome easily.
- **Thermosetting** polymers do not soften when heated because the **cross-links** are **covalent bonds** which lock the polymer chains together and stop them from being flexible.

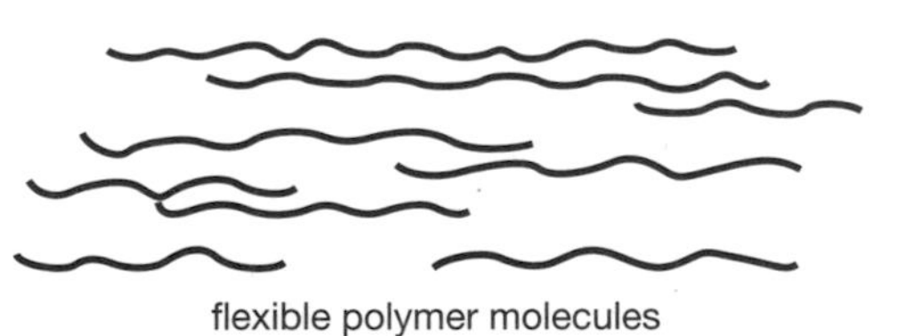

flexible polymer molecules

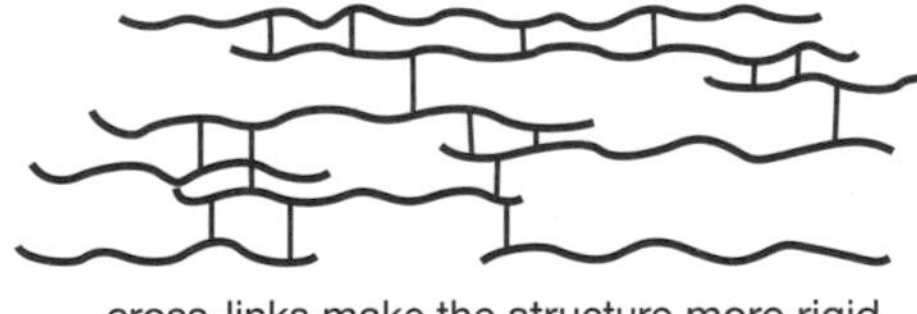

cross-links make the structure more rigid

Thermosoftening polymers are made of long flexible molecules. Thermosetting polymers have cross-links between neighbouring molecules, giving the polymer a three-dimensional rigid structure

examiner tip

Polymers made under different conditions have different properties. Remember that these 'conditions' include the reactants, temperature, pressure and type of catalyst.

Check your understanding

Tested

14 What is an ionic bond? *(4 marks)*

15 Explain the high melting points of ionic compounds. *(4 marks)*

16 Why do ionic compounds conduct electricity when dissolved in water? *(2 marks)*

17 Why do metals conduct electricity? *(2 marks)*

18 Why are alloys harder than pure metals? *(2 marks)*

19 What is a shape memory alloy? *(1 mark)*

20 What determines the property of a polymer? *(1 mark)*

21 What is the effect of introducing cross-links into a polymer? *(1 mark)*

Answers online **Test yourself online** Online

Nanoscience

Nanoscience refers to structures that are **1–100 nm** in size, of the order of only a few hundred atoms. The smallest particle of a material you could slice with the sharpest knife would be 1000 times bigger.

Nanoparticles show different properties from the same material in bulk — mainly because they have a **high surface-area-to-volume ratio**.

These novel properties may lead to the development of:

- new computers
- new catalysts
- new coatings
- highly selective sensors
- stronger and lighter construction materials
- new cosmetics such as suntan creams and deodorants

examiner tip

Nanoparticles are so different from normal because they are so small.

Choosing materials for different uses

Revised

The different properties of materials make them useful for different applications.

You may be provided with information about the properties of substances that are not specified in this unit but you will need to be able to relate the properties to their uses. For example: insert the appropriate uses into a copy of the table below. Choose from this list:

- windows
- ropes
- pipes
- electrical insulation
- foam packaging
- non-stick frying pans
- thermal insulation
- bowls
- carrier bags
- guttering
- soles of irons
- packaging (more than one polymer)
- buckets
- rulers

Polymer	Properties	Uses
Polyethene	Tough yet flexible	
Polypropene	Tough and durable; high tensile strength	
Polychloroethene (PVC)	Strong and hard but less flexible than polyethene	
PTFE	Non-stick and withstands high temperatures	
Polystyrene	Low density and poor conductor of heat	
Perspex	Transparent and rigid	

(The answers are given at the bottom of this page.)

Developments and applications of new materials

Revised

While you need to be familiar with some examples of new materials you do not need to know specific properties or names of them in the examination. However, you may well have to **evaluate** developments and applications of some given examples — perhaps a shape memory metal used in spectacle frames.

Polymer	Uses
Polyethene	Carrier bags, bowls, buckets, packaging
Polypropene	Packaging, ropes
Polychloroethene (PVC)	Pipes, electrical insulation, guttering
PTFE	Non-stick frying pans, soles of irons
Polystyrene	Thermal insulation, foam packaging
Perspex	Windows, rulers

Using information from text provided in the question you can **draw a table** as your answer, like this:

Metal used for the frames	Benefits	Drawbacks
Steel	Hard, tough and strong Alloys are harder than pure metals	High density, therefore heavy to wear Corrodes in contact with perspiration
Nitinol (a shape memory metal)	Can return to its original shape Strong and resists corrosion More flexible than steel Low density, therefore lightweight frames	High cost Increased metal fatigue

examiner tip

Using a table for your answer will help you analyse the data properly and stop you being vague.

And don't worry! Look at some mark schemes online — they are in table form themselves.

Structure and properties

Revised

Another type of question could ask you to suggest the type of structure of a substance given its properties.

Structures and properties that you should know are shown in the table:

Structure	Melting and boiling points	Electrical conductivity	
		When solid	When molten (liquid)
Simple molecular	low	nil	nil
Giant molecular	high	nil	nil
Giant ionic	high	nil	good
Metallic	high	good	good

Check your understanding

Tested

22 How big are nanoparticles? *(1 mark)*

23 Why do nanoparticles have such different properties from materials made of normal-size particles? *(1 mark)*

24 Using data from the table, answer the questions below.

Metal	Density in g/cm³	Relative strength compared with iron	Relative electrical conductivity compared with iron	Relative thermal conductivity compared with iron	Cost per tonne in £
Aluminium	2.7	0.33	3.7	3.0	950
Copper	8.9	0.62	5.8	4.8	1100
Iron	7.9	1.00	1.0	1.0	130
Silver	10.5	0.39	6.1	5.2	250 000

a) Why are the wires between electricity pylons made of aluminium with a steel core? *(3 marks)*

b) Why is copper, but not silver, used to make pots and pans? *(3 marks)*

25 Suggest the type of structure for each substance in the following table. *(4 marks)*

Substance	Boiling point (°C)	Electrical conductivity		Structure
		Solid	Liquid	
A	2730	Good	Good	
B	–78	Poor	Poor	
C	2980	Poor	Good	
D	2230	Poor	Poor	

Answers online — Test yourself online — Online

Atomic structure

Isotopes and atomic mass

Revised

Atoms can be represented as shown alongside.

The relative masses of protons, neutrons and electrons are as follows:

Name of particle	Mass
Proton	1
Neutron	1
Electron	Very small

mass number = 23

Na

atomic number = 11

↑ The symbol for sodium shown with its mass number and atomic number

- The total number of protons and neutrons in an atom is called its **mass number**.
- Atoms of the **same element** can have **different** numbers of **neutrons**; these atoms are called **isotopes** of that element.

Remember that atoms of the same element have the same proton number. Isotopes of an element therefore have different mass numbers — the sum of protons and neutrons is different. For example, $^{24}_{11}Na$ and $^{23}_{11}Na$ are isotopes of sodium with different mass numbers.

Isotopes of an element have the **same chemical properties** — their electronic structures are the same, so they react in the same way — but **different physical properties**, such as density, boiling point and freezing point.

Isotopes are the same element but have different atoms

Isotopes have the same:	Isotopes have different:
number of protons	numbers of neutrons
number of electrons	mass numbers
atomic number	physical properties
chemical properties	

Relative atomic mass

Revised

- The **relative atomic mass** of an element (A_r) compares the mass of atoms of the element with the ^{12}C isotope, which is given the atomic mass of exactly 12.
- It is an average value for the isotopes of the element.

For example, atoms of chlorine in compounds or the element are found either as the isotope ^{35}Cl or as ^{37}Cl; 75% of all the atoms are ^{35}Cl and 25% are ^{37}Cl.

So, the relative atomic mass of chlorine can be calculated as:

$$A_r = \frac{75}{100} \times 35.00 + \frac{25}{100} \times 37.00$$
$$= 26.25 + 9.25$$
$$= 35.50$$

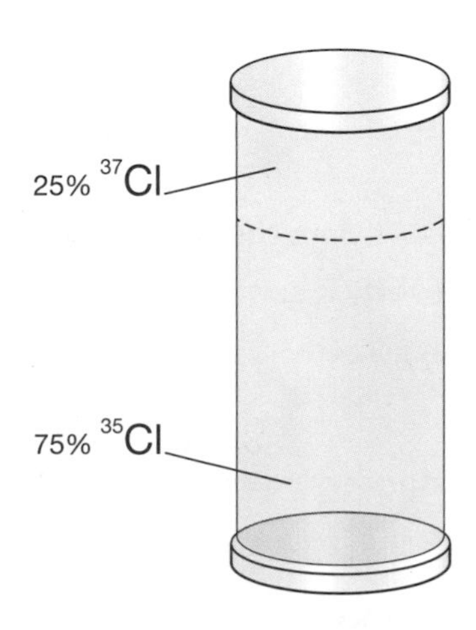

↑ 25% of chlorine atoms have atomic mass 37; 75% of chlorine atoms have atomic mass 35

examiner tip

Learn this method well and know how to apply it in cases where more than two isotopes are involved.

Relative formula mass

Revised

- The **relative formula mass (M_r)** of a compound is the sum of the relative atomic masses of the elements in the numbers shown in the formula.
- So the relative formula mass of calcium carbonate ($CaCO_3$) is calculated as:

$M_r = A_r$ calcium + A_r carbon + (A_r oxygen $\times$ 3)

$= 40 + 12 + (16 \times 3)$

$= 40 + 12 + 48$

$= 100$

- Likewise, the relative formula mass of water (H_2O) is calculated as:

M_r = (A_r hydrogen $\times$ 2) + A_r oxygen

$= (1 \times 2) + 16$

$= 2 + 16$

$= 18$

- The relative formula mass of a substance, in grams, is known as **one mole** of that substance.
- number of moles $= \dfrac{\text{mass of substance}}{\text{relative formula mass}}$

examiner tip

moles $= \dfrac{\text{mass}}{\text{formula mass}}$

Look at the table for some examples.

Moles of materials

Substance	Formula	Type of particles	Formula mass (M_r)	Mass of ½ mole in g	Mass of 1 mole in g	Mass of 2 moles in g	Mass of 5 moles in g
Water	H_2O	Molecules	18	9	18	36	90
Oxygen	O_2	Molecules	32	16	32	64	160
Carbon	C	Atoms	12	6	12	24	60
Methane	CH_4	Molecules	16	8	16	32	80
Sodium chloride	NaCl	Ions	58.5	29.25	58.5	117	292.5

Check your understanding

Tested

26 Explain why different isotopes of an element have different masses. *(2 marks)*

27 How many neutrons are there in an atom of the isotope $^{210}_{84}Po$? *(2 marks)*

28 Calculate the relative atomic masses of the following:

a) Copper, as it occurs naturally, which is 69.0% ^{63}Cu and 31.0% ^{65}Cu. *(2 marks)*

b) Sulfur, as it occurs naturally, which is 95.00% ^{32}S, 0.77% ^{33}S and 4.23% ^{34}S. *(2 marks)*

29 What is the relative formula mass of each of the following substances?

a) NaOH *(1 mark)*

b) Cl_2 *(1 mark)*

c) H_2SO_4 *(1 mark)*

d) $Fe_2(SO_4)_3$ *(1 mark)*

Answers online **Test yourself online** Online

Analysing substances

- Elements and compounds can be **detected** and **identified** using **instrumental** methods.
- **Instrumental** methods are accurate, sensitive and rapid and are particularly useful when a sample is very **small**.

The instrumental methods you have probably come across will have mostly been various types of spectroscopy.

They are used for many purposes but the more well known ones are the detection and identification of illegal drugs and explosives.

- **Chemical analysis** can be used to identify additives in foods.

Chemical analysis of foods involves reacting food samples with chemicals that produce known results when certain chemicals are present in the food. This can be slow, laborious and inaccurate.

Paper chromatography

Revised

- **Artificial colours** can be detected and identified by **paper chromatography**. You will probably have done this with inks — from bottles or using the tips of felt pens. You may have even used food colourings.

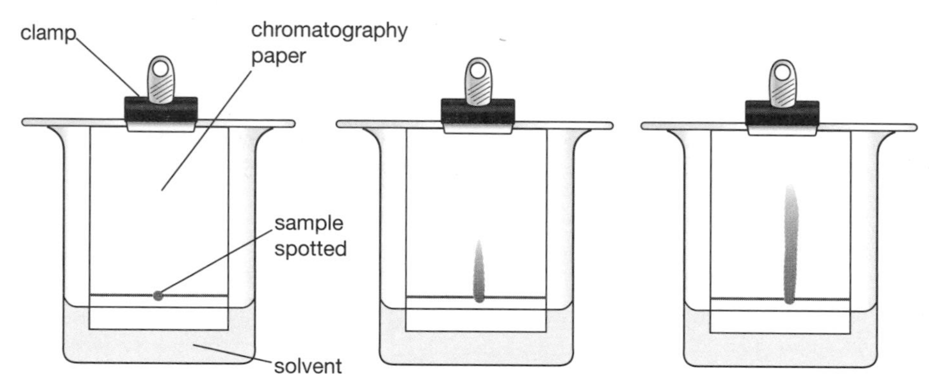

Simple paper chromatography

- A pencil line is drawn near the bottom of the paper and marked where the test spots are to be placed.
- Small spots of different materials are applied to the pencil marks.
- The paper is carefully suspended in a solvent, ensuring that the spots are above its surface.
- The paper is removed from the solvent before the solvent front has reached the top.
- Chromatography works because the different molecules are attracted to other substances with different strengths and have different solubilities.
- Therefore a given pure colour will travel a given distance up the chromatography paper and be separated from the other colours.
- The diagram alongside shows that the orange spot contains red and green colours.

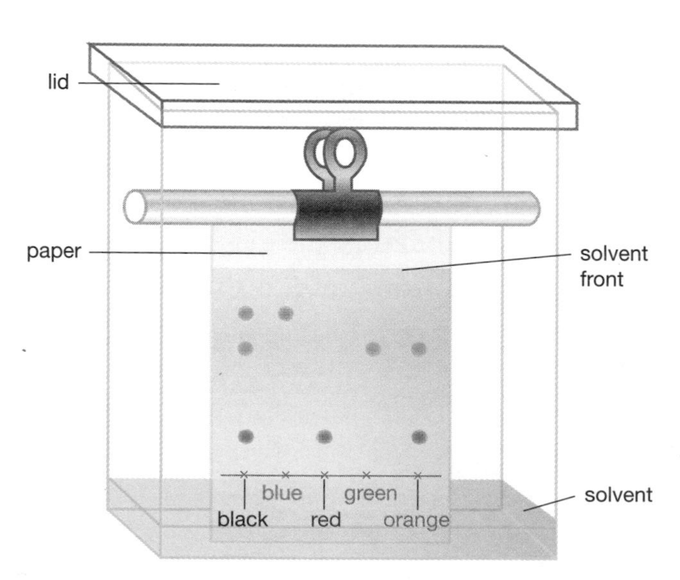

Separation of colours using paper chromatography

examiner tip

Use a ruler held across any chromatogram to accurately find if spots from different starting colours have travelled the same distance up the paper.

Gas chromatography and mass spectroscopy

Revised

Gas chromatography linked to mass spectroscopy (GC-MS) is an example of an instrumental method.

- **Gas chromatography** allows the **separation** of a mixture of compounds.
- The **time taken** for a substance to travel through the column can be used to **help identify** the substance. This is known as the **retention time**.

This is how it works:

- The sample is vaporised and injected into a long (coiled) tube. This is the **column**, and it is packed with a solid powdered material.
- The sample is carried by an unreactive gas through the column.
- Different substances in the sample are carried at different speeds, so they become separated.
- The components of the sample are detected as they come out of the other end; a **chromatogram** is produced.
- The **number of peaks** on the chromatogram shows the **number of compounds** present.
- The **positions** of the peaks on the chromatogram indicate the **retention times**.
- The output from the gas chromatography column can be linked to a **mass spectrometer**, which then identifies the substances leaving the column.
- A **mass spectrometer** can identify substances very quickly and accurately and can detect very small quantities (sometimes as small as 10^{-18} mole).

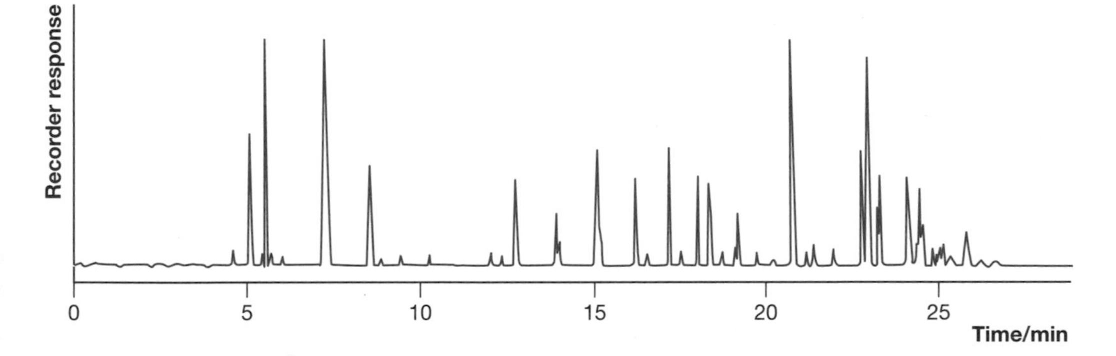

← A gas chromatogram of a sample of petrol — a mixture of hydrocarbons

- The mass spectrometer can also give the **relative molecular mass** of each of the substances separated in the column. This is given by the **molecular ion** peak in the mass spectrum.
- The **molecular ion** peak is the short peak at the far right of a mass spectrum.

Intensity 100% — peaks labelled 15, 29, 43, 58; Mass axis 0 10 20 30 40 50 60 70

→ The mass spectrum of a hydrocarbon. Note the molecular ion peak at mass 58

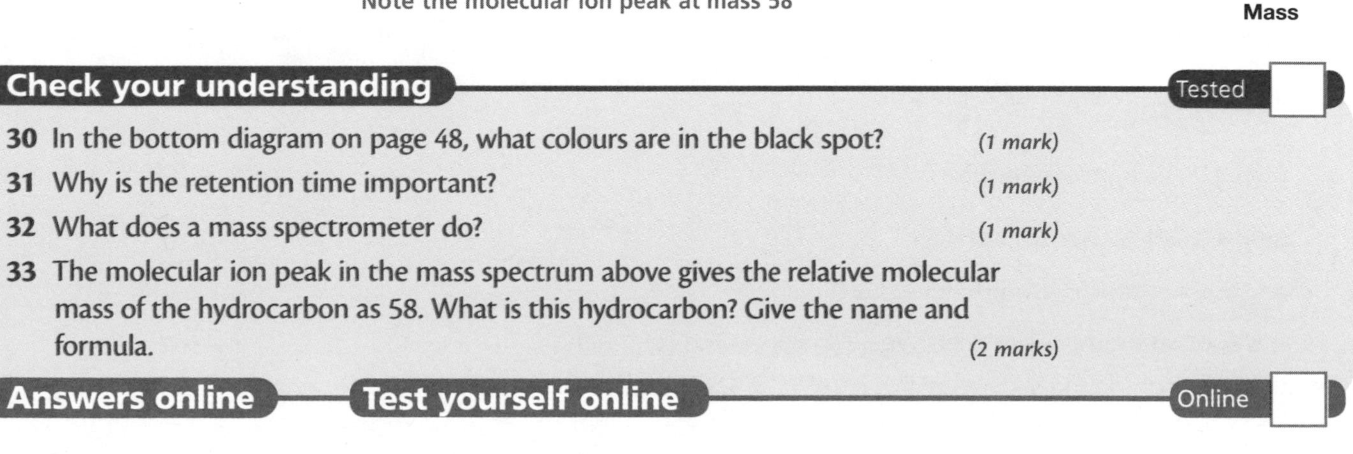

Check your understanding

Tested

30 In the bottom diagram on page 48, what colours are in the black spot? *(1 mark)*

31 Why is the retention time important? *(1 mark)*

32 What does a mass spectrometer do? *(1 mark)*

33 The molecular ion peak in the mass spectrum above gives the relative molecular mass of the hydrocarbon as 58. What is this hydrocarbon? Give the name and formula. *(2 marks)*

Answers online **Test yourself online** Online

Quantitative chemistry

Percentage composition of compounds

Revised

- The **percentage** of an element in a compound can be calculated from **the relative atomic mass** of the element in the formula and **the relative formula mass** of the compound.

This is a lot easier to work out than it sounds.

Example 1: Carbon dioxide (CO_2)

Its **formula mass** (M_r) is 44. Carbon has a relative atomic mass of 12, and that of oxygen is 16 (note this is of the **atom**).

Therefore, carbon dioxide is $\frac{12}{44}$ parts carbon and $2 \times \frac{16}{44}$ parts oxygen.

Changing to percentages it looks like this:

$$\frac{12}{44} \times 100 = 27\% \text{ carbon}$$

$$2 \times \frac{16}{44} \times 100 = 73\% \text{ oxygen}$$

Example 2: It is important to know the percentage of nitrogen in fertiliser.

Ammonium nitrate, NH_4NO_3, is an important ingredient of many fertilisers.

Calculate the percentage of nitrogen in this fertiliser.

NH_4NO_3: $M_r = 80$

percentage nitrogen $= 2 \times \frac{14}{80} \times 100 = 35\%$

Empirical formulae

Revised

- When chemists work out the formula of a compound from experimental results it is called the **empirical formula**.

Suppose you are told that analysis of 72 g of a copper oxide showed it consists of 64 g copper.

In this case you have first to work out the mass of the sample that is due to oxygen (you are told that it is an oxide).

mass of oxygen = total mass – mass of copper
= 72 g – 64 g
= 8 g

Now make a column for both elements and follow the steps.

What to do	Elements	
Element symbol	Cu	O
Step 1 Mass or percentage	64	8
Step 2 Divide by relative atomic mass	$\frac{64}{64} = 1$	$\frac{8}{16} = 0.5$
Step 3 Divide by smallest number	$\frac{1}{0.5}$	$\frac{0.5}{0.5}$
Step 4 Smallest whole number ratio	2	1

And to finish off — write the formula: Cu_2O

Here is another example, but with three elements in the compound.

Analysis of a compound showed that it consists of 32% copper, 6% carbon and 24% oxygen. Make a column for all three elements and follow the steps as before.

What to do	Elements		
Element symbol	Cu	C	O
Step 1 Mass or percentage	32	6	24
Step 2 Divide by relative atomic mass	$\frac{32}{64} = 0.5$	$\frac{6}{12} = 0.5$	$\frac{24}{16} = 1.5$
Step 3 Divide by smallest number	$\frac{0.5}{0.5}$	$\frac{0.5}{0.5}$	$\frac{1.5}{0.5}$
Step 4 Smallest whole number ratio	1	1	3

And to finish off — write the formula: $CuCO_3$

Calculating amounts in reactions

Revised

- The masses of reactants and products can be calculated from **balanced symbol equations**.

Consider the reaction between iron(III) oxide and carbon monoxide, producing iron (as happens in the blast furnace). How much iron will be produced from 1 kg of iron(III) oxide?

Step 1 You have to start with the **balanced equation**:

$Fe_2O_3 + 3CO \rightarrow 2Fe + 3CO_2$

This shows the proportions of the reacting masses, in moles.

Now write down the following, under the symbols of the products and reactants you are interested in:

Step 2 Relative formula mass in grams:

$160\,g\ Fe_2O_3 \rightarrow 2 \times 56\,g\ Fe$

Work out how much 1 g of the oxide would produce:

Step 3 Divide both sides by 160:

$$\frac{160\,g\ Fe_2O_3}{160} \rightarrow \frac{112\,g\ Fe}{160}$$

$1\,g\ Fe_2O_3 \rightarrow 0.7\,g\ Fe$

Now working out how much 1 kg of the oxide would produce:

Step 4 Multiply both sides by 1000:

$1\,kg\ Fe_2O_3 \rightarrow 700\,g\ Fe$

You may have to do this the other way round as well — how much iron(III) oxide is needed to produce 1 kg of iron?

Showing the **four steps** we have:

Step 1 Balanced equation	$Fe_2O_3 + 3CO \rightarrow 2Fe + 3CO_2$
Step 2 Relative formula mass in grams	$160\,g\ Fe_2O_3 \rightarrow 2 \times 56\,g\ Fe$
Step 3 Divide both sides by 112	$\frac{160\,g\ Fe_2O_3}{112} \rightarrow \frac{112\,g\ Fe}{112}$
	$1.4\,g\ Fe_2O_3 \rightarrow 1\,g\ Fe$
Step 4 Multiply both sides by 1000	$1.4\,kg\ Fe_2O_3 \rightarrow 1\,kg\ Fe$

Yields

Revised

Although **no atoms are gained or lost in a chemical reaction** it is not always possible to obtain the calculated amount of a product for the following reasons:

1 The reaction may not go to completion because it is **reversible**. There is always some reactant in an equilibrium mixture — it has not reacted to make the product.

2 Some of the product may be **lost** when it is separated from the reaction mixture. Yield is often reduced when the crude product is washed or distilled, or passed through a filtration process — it can be left on the filter or may be washed through and discarded.

3 Some of the reactants can **react in ways that are different** from the expected reaction. **By-products** are formed in many industrial reactions and processes.

examiner tip

Learn these three reasons and list them if you are asked to suggest reasons why the yield is not 100% in a chemical reaction.

The amount of a product obtained is known as the **yield**. When compared with the maximum theoretical amount as a percentage, it is called the **percentage yield**.

$$\text{percentage yield} = \frac{\text{actual mass of product}}{\text{theoretical mass of product}} \times 100\%$$

For example, if a chemical reaction actually produces 20 g of product when in theory it should produce 40 g then:

$$\text{percentage yield} = \frac{20\,\text{g}}{40\,\text{g}} \times 100\% = 50\%$$

Reversible reactions

Revised

- In some chemical reactions, the products of the reaction react to produce the original reactants.
- Such reactions are called **reversible reactions** and are represented as:

$A + B \rightleftharpoons C + D$

For example:

ammonium chloride $\rightleftharpoons$ ammonia + hydrogen chloride

$NH_4Cl \rightleftharpoons NH_3 + HCl$

reactants $\rightleftharpoons$ products

↑ **This sign shows that the reaction can take place in both the forward and the reverse directions, depending on the conditions used**

Check your understanding

Tested

34 What is the percentage of iron in iron oxide (Fe_2O_3)? *(3 marks)*

35 Some copper oxide (8.0 g) was found to contain 6.4 g copper. What is its empirical formula? *(3 marks)*

36 What mass of ethanol (C_2H_5OH) will be produced by fermenting 100 g of sugar ($C_6H_{12}O_6$), given that the balanced equation is:

$C_6H_{12}O_6 \rightarrow 2C_2H_5OH + 2CO_2$ *(3 marks)*

37 What is the yield of a certain reaction that results in 4.8 g of product when 6.0 g is expected from the balanced equation? *(1 mark)*

38 Why do some chemical reactions produce less than 100% yield? *(3 marks)*

Answers online **Test yourself online** Online

Rates of reaction

- The rate of a reaction can be found by measuring the amount of a reactant used or the amount of product formed over time:

$$\text{rate of reaction} = \frac{\text{amount of reactant used}}{\text{time}} \quad \text{or} \quad \text{rate of reaction} = \frac{\text{amount of product formed}}{\text{time}}$$

- This is simply either how fast a reactant is being used or a product is being formed.

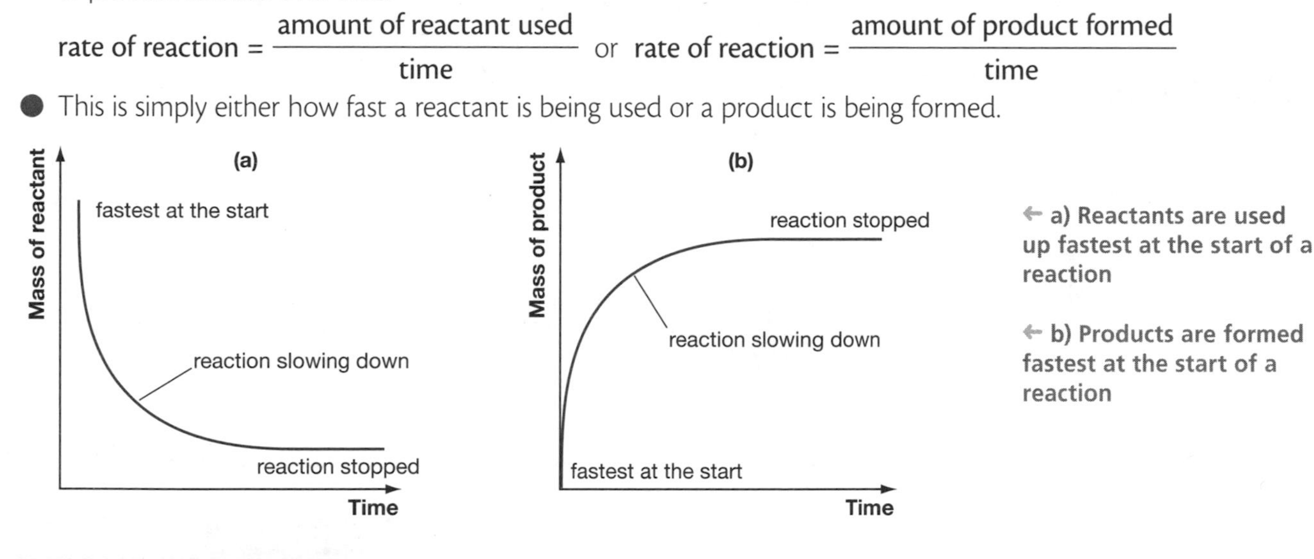

← a) Reactants are used up fastest at the start of a reaction

← b) Products are formed fastest at the start of a reaction

Collision theory

Revised

- Chemical reactions can occur only when reacting particles **collide** with each other and with **sufficient energy**.
- The minimum amount of energy particles must have to react is called the **activation energy** for the reaction.

Therefore, not all collisions are 'successful' i.e. not all collisions result in a reaction between the particles. The rate of a reaction depends on the number of successful collisions in a given time.

↑ A reaction can occur only if the collision transfers enough energy

Increasing rates of reaction

Revised

Increasing the **temperature increases**:

- the **speed** of the reacting particles
- the **frequency** of collision
- the **energy** of collision
- and therefore the **rate** of reaction

→ Increasing temperature increases the rate of reaction

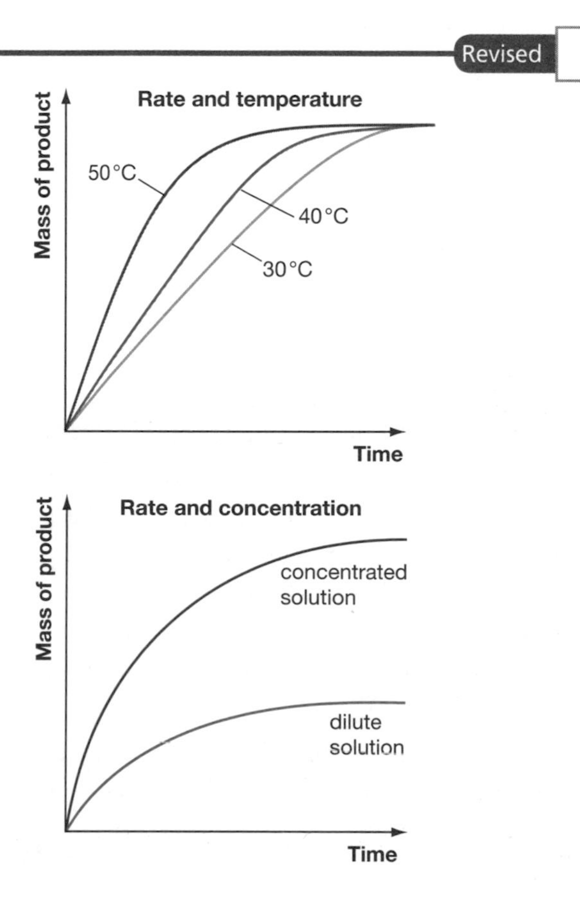

Increasing the **pressure** of reacting gases **increases**:

- the **frequency** of collisions
- and therefore the **rate** of reaction

Increasing the **concentration** of reactants in solution **increases**:

- the **frequency** of collisions
- and therefore the **rate** of reaction

→ Increasing the concentration or pressure of reactants increases the rate of reaction

Increasing the **surface area** of solid reactants **increases**:

- the **frequency** of collision
- and therefore the **rate** of reaction

→ **Increasing surface area increases the rate of reaction**

Rate and particle size

Mass of product

small particles large surface area

large particles small surface area

Time

examiner tip

The rate of a reaction increases for just **one** reason: the number of successful collisions increases.

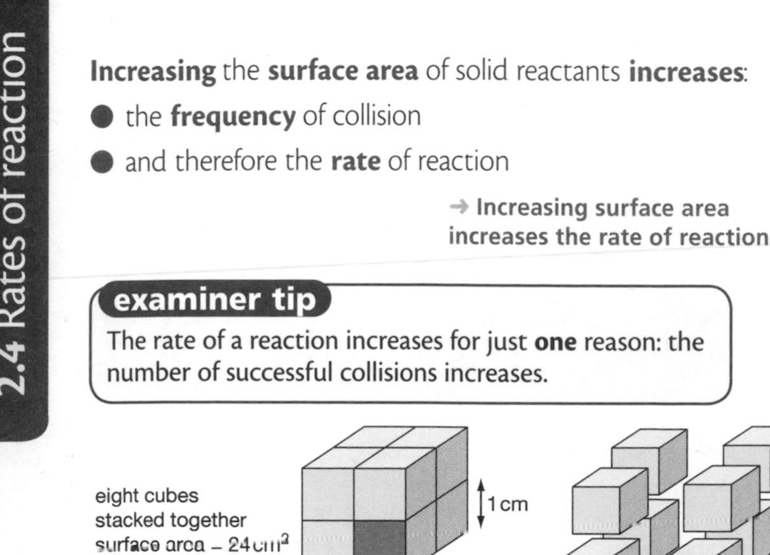

← **Smaller pieces increase the surface area and the number of particles exposed on the surface**

Catalysts

Revised

- **Catalysts** change the rate of chemical reactions but are not used up during the reaction — they can be used over and over again.
- **Different reactions** need **different catalysts**.
- Catalysts are important in **increasing the rates** of chemical reactions used in industrial processes to **reduce costs**.

(The idea of catalysts reducing the activation energy of a reaction is explored on page 79.)

examiner tip

You will not have to know specific examples of processes and their catalysts.

You will be given information to evaluate — use it all and consider using a table for your answer.

Advantages and disadvantages of using catalysts

Revised

Reasons for developing a catalyst for an industrial process include:

- It speeds up the reaction, reducing the time needed to produce a given amount.
- It reduces the energy costs involved.
- It reduces the labour (time) costs involved.

Advantages of using a catalyst include all the above, plus:

- The reaction can be done at a lower temperature, which is safer as well as cheaper.
- It is not used up, so can be used many times.

Some disadvantages of using a catalyst include:

- It may be an expensive material (e.g. platinum).
- It may be difficult to remove from the product.

Check your understanding

Tested

39 In the first diagram, when is the reaction fastest? *(1 mark)*

40 The first graph shows the mass of CO_2 lost as hydrochloric acid reacted with marble chips. What mass of gas was lost between 10 and 40 minutes? *(1 mark)*

41 After what time had the reaction finished? *(1 mark)*

42 In the second diagram, which reaction was the fastest? *(1 mark)*

43 Which reaction in the second diagram had the greatest amount of starting materials? *(1 mark)*

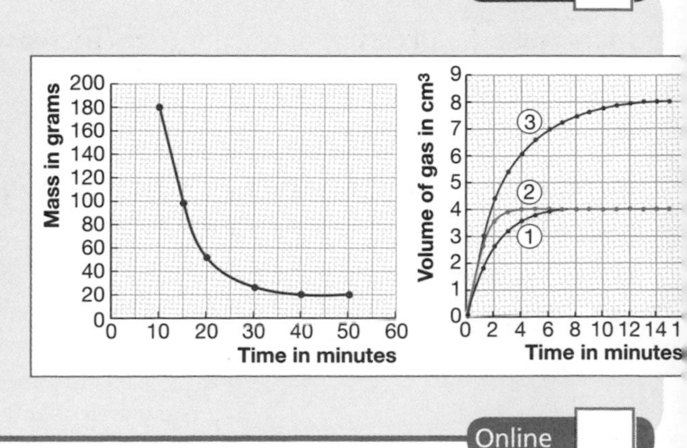

Answers online **Test yourself online**

Online

Energy transfer in chemical reactions

Exothermic and endothermic

Revised

- When chemical reactions occur, **energy** is **transferred to** or **from** the **surroundings**.
- In **exothermic** reactions **energy** is transferred **to** the surroundings. This usually causes the surroundings to get **hotter**.

Remember: the surroundings may well be the water of the solution where the reaction is happening. The water has energy transferred to it from the chemicals that are reacting in it, so it gets hot.

Exothermic reactions include:

- **combustion** (e.g. burning methane in a Bunsen burner)
- **many oxidation reactions** (e.g. copper turning to copper oxide)
- **neutralisation** (acid and alkali get warm when mixed).

> **examiner tip**
>
> Keep it simple:
> - exothermic reactions get **hot**
> - endothermic reactions get **cold**

Uses of exothermic reactions include **self-heating cans** (e.g. for soup) and **hand warmers**.

In these examples of exothermic reactions, the energy transferred is **heat** energy. **Light** energy can also be transferred to the surroundings, in which case the reaction will produce light. Glow sticks gently **glow** (and they **do not** get hot) but fireworks burn **brightly** (and **do** get very hot).

- In **endothermic** reactions **energy** is taken in **from** the surroundings. This causes the surroundings to get **colder**.
- Endothermic reactions include **thermal decomposition**.

You came across a thermal decomposition when you looked at calcium carbonate (probably as a marble chip) being heated to form calcium oxide and carbon dioxide. Although the marble chip is hot, it is **absorbing** heat from the Bunsen flame.

An endothermic reaction occurs when you pour salt onto ice. This produces a remarkable drop in temperature because the reaction takes energy from the surroundings, which is the **ice**, so it **gets colder**. (This is not to be confused with what happens when you put salt on an icy path in winter. Here the salt causing water to have a lower freezing point is the important factor.)

Some **sports injury packs** are based on endothermic reactions.

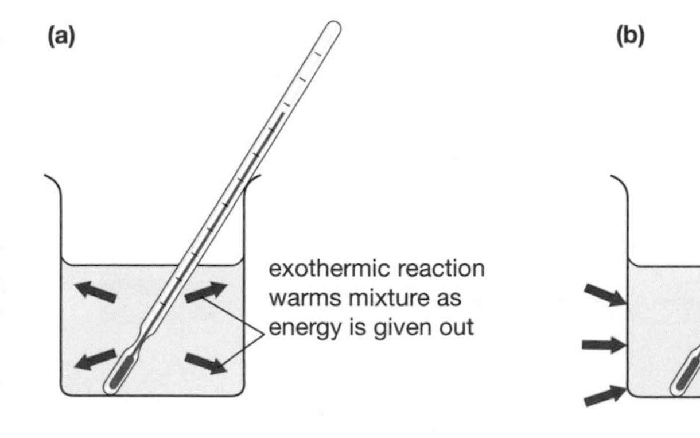

a) An exothermic reaction transfers energy to the surroundings

b) An endothermic reaction absorbs energy from the surroundings

Energy changes in reversible reactions

Revised

- **Reversible reactions** are introduced on page 52.
- If a **reversible reaction** is exothermic in one direction, it is endothermic in the **opposite** direction.

- The **same** amount of **energy** is transferred in each case for the same amount of reactants.
- An example of energy changes in a reversible reaction is the dehydration of copper sulfate ($CuSO_4.5H_2O$).
- If copper sulfate is heated in a test tube, water is driven off. This is seen as steam. The copper sulfate turns from blue to white as it absorbs energy from the heat source.

hydrated copper sulfate (blue) $\underset{\text{exothermic}}{\overset{\text{endothermic}}{\rightleftharpoons}}$ **anhydrous copper sulfate** (white) + **water**

$$CuSO_4.5H_2O \underset{\text{exothermic}}{\overset{\text{endothermic}}{\rightleftharpoons}} CuSO_4 + 5H_2O$$

- When, after cooling back to room temperature, water is put onto the white powder a lot of heat is given out and the copper sulfate gets hot as it returns to its original blue colour.

Test for water

Revised

The reverse copper sulfate reaction can be used as a **test for water**:

- If it is thought that a liquid contains water or is only water, a drop or two can be added to some anhydrous (white) copper sulfate powder or paper treated with it.
- If water is present, the copper sulfate turns blue.

In reality, anhydrous copper sulfate has a blue tinge due to water already absorbed from the air.

Good uses of energy changes

Revised

Before an exothermic or endothermic reaction can be put to use around the home it has to be shown to be effective and safe. In other words it has to be **evaluated**.

The everyday examples above include self-heating cans for soup, hand warmers, glow sticks and sports injury packs.

You could be given data from which to draw conclusions. This might include information such as:

- the maximum or minimum temperature attained by the reaction
- how long it stayed hot or cold
- the chemical hazards involved
- how can the product be disposed of

Remember to give a **balanced** argument — include **advantages** and **disadvantages** and make sure you **use the data** from the question.

Check your understanding

Tested

44 Describe the energy transfer in an exothermic reaction. *(1 mark)*

45 Describe the energy transfer in an endothermic reaction. *(1 mark)*

46 Give three examples of exothermic reactions. *(3 marks)*

47 Describe what happens when water is added to anhydrous (white) copper sulfate. *(2 marks)*

Answers online **Test yourself online** Online

Making salts

Soluble salts

Revised

Soluble salts can be made from **acids** by reacting them with:

- **metals**
- **insoluble bases**
- **alkalis**

Metals: Not all metals are suitable; some are too reactive (you are unlikely to see sodium reacting with an acid) and others (like gold) are not reactive enough.

Insoluble bases: A base such as copper oxide is added to the acid until no more will react (dissolve). This is when there is unreacted solid at the bottom of the mixture. This **excess solid** is **filtered** off.

Alkalis: An **indicator** can be used to show when the acid and alkali have completely reacted to produce a **salt solution**.

All these reactions with acids are **neutralisation**.

Salt solutions can be **crystallised** to produce solid salts. Preparation of a solid salt follows the four steps shown in the diagrams.

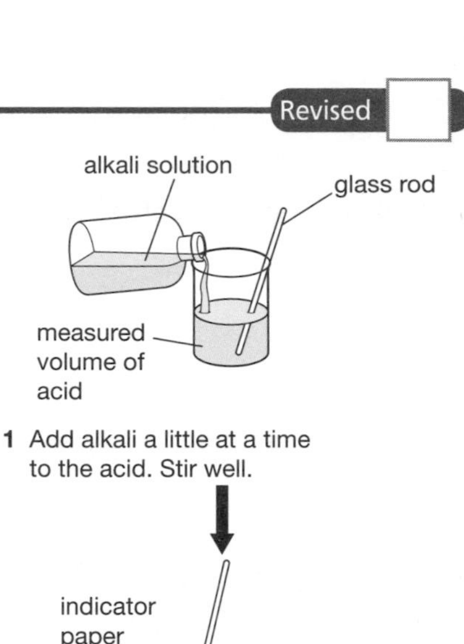

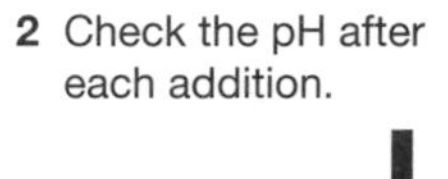

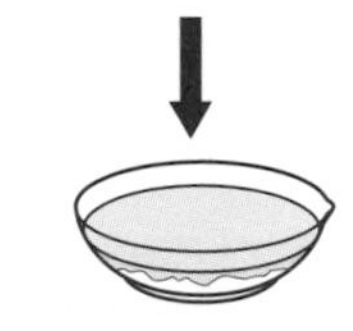

↑ **Preparing a salt by neutralisation**

examiner tip

Make sure you know how to obtain a salt from a solution.

Crystallisation is always the same — evaporate to crystallising point then set aside to cool and form crystals.

The general equations for the three methods are as follows:

acid + metal → salt + hydrogen

acid + base → salt + water

acid + alkali → salt + water

The word equations for the neutralisation of **hydrochloric acid** using each of the methods are:

- **metals**: hydrochloric acid + zinc → zinc chloride + hydrogen
- **insoluble bases**: hydrochloric acid + copper oxide → copper chloride + water
- **alkalis**: hydrochloric acid + sodium hydroxide → sodium chloride + water

The tables on page 60 tell you how the **name** of the salt depends on the acid used.

- The **state symbols** in equations are **(s)** for solid, **(l)** for liquid, **(g)** for gas and **(aq)** for aqueous (water) solution.
- Metals and insoluble bases will be solids **(s)**.
- The acids used and salt solutions produced will be aqueous **(aq)**.
- Water is a liquid **(l)**.

The balanced equations for these examples are:

$2HCl(aq) + Zn(s) \rightarrow ZnCl_2(aq) + H_2(g)$

$2HCl(aq) + CuO(s) \rightarrow CuCl_2(aq) + H_2O(l)$

$HCl(aq) + NaOH(aq) \rightarrow NaCl(aq) + H_2O(l)$

examiner tip

A liquid is not a solution. A liquid is a pure substance; a solution is a solvent with a solute dissolved in it.

Insoluble salts

Revised

- Insoluble salts can be made by mixing appropriate solutions of ions so that a **precipitate** is formed.

This table gives some examples of solutions that could be mixed to make the named insoluble salts.

Insoluble salt	Solution 1	Solution 2
Silver chloride	Silver nitrate	Sodium chloride
Barium sulfate	Barium nitrate	Sodium sulfate
Lead iodide	Lead nitrate	Sodium iodide

All **nitrates** are soluble — you can always use a **metal nitrate** to provide the metal part of the salt.

Precipitation and water treatment

Revised

- Precipitation can be used to remove **unwanted ions** from solutions.
- The precipitate is the **insoluble compound** and can be filtered from the mixture.
- When an insoluble compound is made, the **ions** that were **free to move** in the separate solutions become **locked up** in the giant ionic structure; they have been **removed from solution**.

For example: calcium hydroxide solution (limewater) can be used to:

1 remove iron (Fe^{3+}(aq)) ions from **drinking water**:

$$2Fe^{3+}(aq) + 3Ca(OH)_2(aq) \rightarrow 2Fe(OH)_3(s) + 3Ca^{2+}(aq)$$

The solid iron(III) hydroxide is filtered off.

2 remove heavy metal ions such as cadmium and lead from **industrial effluent**:

$$2Pb^{2+}(aq) + 2Ca(OH)_2(aq) \rightarrow 2Pb(OH)_2(s) + 2Ca^{2+}(aq)$$

Again, the solid is filtered off.

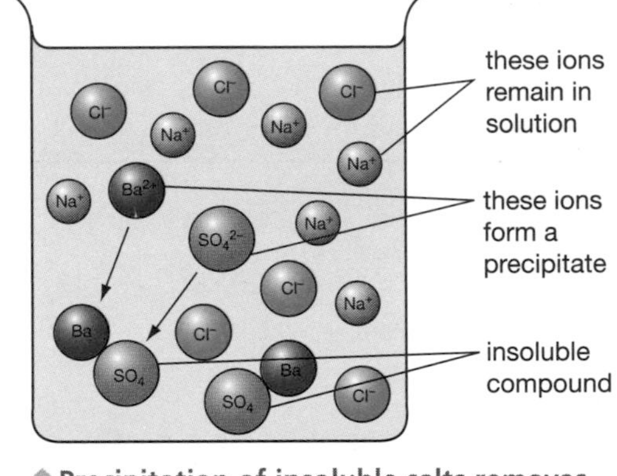

↑ **Precipitation of insoluble salts removes ions from solution**

Check your understanding

Tested

48 Describe how you would prepare a sample of magnesium chloride (a soluble salt) starting with solid magnesium hydroxide and hydrochloric acid. *(4 marks)*

49 Copy and complete these word equations:

a) sodium hydroxide + _________ acid → sodium chloride + water *(1 mark)*

b) copper oxide + nitric acid → _________ + _________ *(2 marks)*

c) sodium hydroxide + sulfuric acid → _________ + _________ *(2 marks)*

d) lead nitrate + potassium chloride → _________ + _________ *(2 marks)*

50 a) How would the addition of limewater to industrial effluent remove dissolved cadmium (Cd^{2+}(aq))? *(2 marks)*

b) Use a balanced chemical equation to help explain your answer. *(4 marks)*

Answers online **Test yourself online** Online

Acids and bases

Acids

Revised

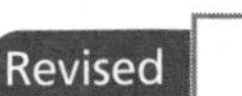

- All acid solutions contain hydrogen ions.
- It is the **hydrogen ions ($H^+(aq)$)** that make solutions **acidic**.

For example, all these are **acidic** because they have **hydrogen ions** in solution:

hydrochloric acid: $HCl(aq) \rightarrow H^+(aq) + Cl^-(aq)$
nitric acid: $HNO_3(aq) \rightarrow H^+(aq) + NO_3^-(aq)$
sulfuric acid: $H_2SO_4(aq) \rightarrow 2H^+(aq) + SO_4^{2-}(aq)$

Alkalis and bases

Revised

- **Metal oxides** and **hydroxides** are **bases**.
- Soluble hydroxides are called **alkalis**.
- It is the **hydroxide ions ($OH^-(aq)$)** that make solutions **alkaline**.

For example, all these are **alkaline** because they have **hydroxide ions** in solution:

sodium hydroxide: $NaOH(aq) \rightarrow Na^+(aq) + OH^-(aq)$
potassium hydroxide: $KOH(aq) \rightarrow K^+(aq) + OH^-(aq)$
calcium hydroxide: $Ca(OH)_2(aq) \rightarrow Ca^{2+}(aq) + 2OH^-(aq)$

examiner tip

Acids are acidic. Alkalis are alkaline.

This table shows a selection of commonly used alkalis and bases.

Alkali (soluble)	Base (insoluble)
Sodium hydroxide	Copper oxide
Potassium hydroxide	Lead oxide
Barium hydroxide	Magnesium hydroxide
	Copper hydroxide

- **Ammonia** is also an **alkali**.
- It dissolves in water to produce an **alkaline solution**:

  $NH_3(g) + H_2O(l) \rightleftharpoons NH_4^+(aq) + OH^-(aq)$
- **Ammonia** is used to produce **ammonium** salts. For example:

  nitric acid + ammonia solution → ammonium nitrate

  $HNO_3(aq) + NH_3(aq) \rightarrow NH_4NO_3(aq)$
- Ammonium salts such as **ammonium nitrate** and **ammonium phosphate** are important as **fertilisers**.
- They are readily soluble in water and can be taken up by the plants to increase growth.

Salts

Revised

- The particular **salt** produced in any reaction between an **acid** and a **base** or **alkali** depends on the **acid** used and the **metal** in the **base** or **alkali**.

Acid used	Salt produced
Hydrochloric	Chloride
Nitric	Nitrate
Sulfuric	Sulfate

Metal in base or alkali used	Metal in salt produced
Copper	Copper
Sodium	Sodium
Potassium	Potassium

By using these two tables together with the one showing alkalis and bases you can suggest which acid you need to react with which alkali or base to produce a named salt.

This table gives some example of acids and alkalis or bases that could be reacted to make the named salts.

Named salt	Acid	Alkali or base
Calcium chloride	Hydrochloric acid	Calcium hydroxide
Potassium nitrate	Nitric acid	Potassium hydroxide
Copper sulfate	Sulfuric acid	Copper oxide
Ammonium phosphate	Phosphoric acid	Ammonia

The pH scale

Revised

- The **pH scale** gives a measure of the acidity or alkalinity of a solution.
- The pH scale runs from 0 to 14:
  – pH 0 is very **acidic**
  – pH 7 is **neutral**
  – pH 14 is very **alkaline**
- **Indicators** are used to show if a substance in solution is **acidic** or **alkaline**.
- **Litmus** only shows if the substance contains an acid or an alkali.
- **Universal indicator** gives more information — it measures the pH, i.e. how acidic or alkaline the material is.
- With universal indicator:
  – pH 1 is deep red
  – pH 7 is green
  – pH 14 is very dark blue

examiner tip

When pH changes during a reaction, look for hydrogen ions or hydroxide ions on the product side:

- H^+ ions decrease the pH.
- OH^- ions increase the pH.

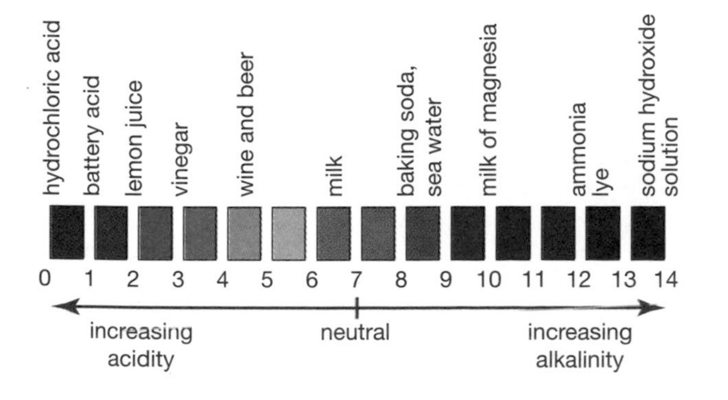

← **Universal indicator paper measures the pH of a solution**

Neutralisation

Revised

- Acids react with alkalis (or bases) in neutralisation reactions:

  acid + alkali → salt + water

  acid + base → salt + water

For example:

nitric acid + sodium hydroxide → sodium nitrate + water

$HNO_3(aq) + NaOH(aq) \rightarrow NaNO_3(aq) + H_2O(l)$

Remember that in solution, nitric acid and sodium hydroxide produce ions like this:

$HNO_3(aq) \rightarrow H^+(aq) + NO_3^-(aq)$

$NaOH(aq) \rightarrow Na^+(aq) + OH^-(aq)$

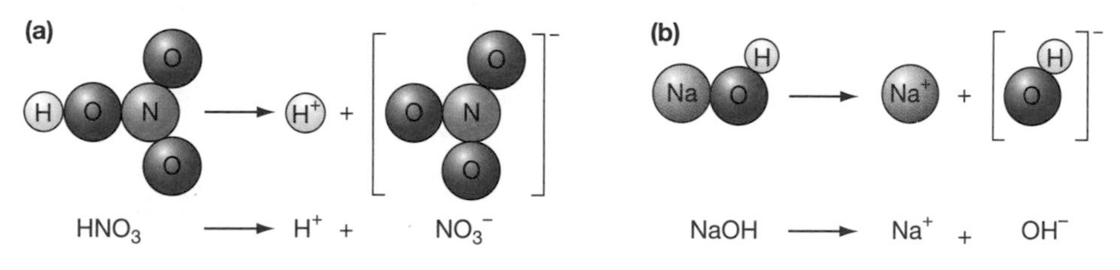

a) Nitric acid splits into hydrogen ions and nitrate ions

b) Sodium hydroxide splits into sodium ions and hydroxide ions

We can now re-write the first equation showing all the separate ions (freely moving in solution):

$H^+(aq) + NO_3^-(aq) + Na^+(aq) + OH^-(aq) \rightarrow Na^+(aq) + NO_3^-(aq) + H_2O(l)$

There are some ions that appear on **both** sides of the equation **in the same state** **(Na^+ and NO_3^-)**.

These are **spectator ions** and take no part in the reaction and stay in solution. They can be cancelled from the equation:

$H^+(aq) + \cancel{NO_3^-(aq)} + \cancel{Na^+(aq)} + OH^-(aq) \rightarrow \cancel{Na^+(aq)} + \cancel{NO_3^-(aq)} + H_2O(l)$

This leaves a much simpler equation for neutralisation between an acid and an alkali:

$H^+(aq) + OH^-(aq) \rightarrow H_2O(l)$

This is the case for any acid reacting with any alkali and so:

- In **neutralisation** reactions, **hydrogen ions** react with **hydroxide ions** to produce **water**.
- This is represented by the equation $H^+ (aq) + OH^-(aq) \rightarrow H_2O(l)$.

examiner tip

Learn these two equations:

- acid + alkali → salt + water
- $H^+(aq) + OH^-(aq) \rightarrow H_2O(l)$.

Check your understanding

Tested

51 What makes acids acidic? *(1 mark)*

52 What makes alkalis alkaline? *(1 mark)*

53 What is the pH value of all neutral solutions? *(1 mark)*

54 What is the equation that most simply represents the neutralisation of hydrochloric acid with sodium hydroxide? *(1 mark)*

55 Give the name of the salt produced when the following compounds react:

a) copper oxide with sulfuric acid *(1 mark)*

b) calcium hydroxide with hydrochloric acid *(1 mark)*

c) magnesium carbonate with nitric acid *(1 mark)*

Answers online **Test yourself online**

Online

Electrolysis (1)

- **Electrolysis** is the conduction of an electric current through **ionic** substances and **breaking them down** into **elements**.
- **Ionic** substances will **conduct electricity** only when **molten** or **dissolved in water**. (See page 42.)
- Only when an ionic substance is **melted** or **dissolved in water** are the ions **free to move about** (mobile) within the liquid or solution.
- The substance that is broken down (**decomposed**) is called the **electrolyte**.

Lead bromide ($PbBr_2$) is used as an **electrolyte** to demonstrate electrolysis:

- it is **ionic**
- it readily **dissolves in water**
- it **melts** relatively easily (about 373°C)

d.c. power supply
carbon electrode
positive lead ions attracted to the negative electrode
molten lead(II) bromide
negative bromide ions attracted to the positive electrode

↑ **Molten lead bromide can be split into its elements (decomposed) by electrolysis**

Electrode reactions

Revised

- Electrolysis involves **chemical reactions** that occur at the electrodes — electrons are lost or gained.
- During electrolysis, positively charged ions move to the negative electrode; negatively charged ions move to the positive electrode.
- At the **negative electrode** (cathode), **positively charged** ions **gain** electrons (**reduction**).
- At the **positive electrode** (anode), **negatively charged** ions **lose** electrons (**oxidation**).
- **Oxidation** or **reduction** does not have to involve **oxygen**.

Electrode reactions produce elements from their ions:

- **Metals** and **hydrogen** are formed at the **cathode**.
- **Oxygen** and the **halogens** (Cl_2, Br_2 and I_2) are formed at the **anode**.

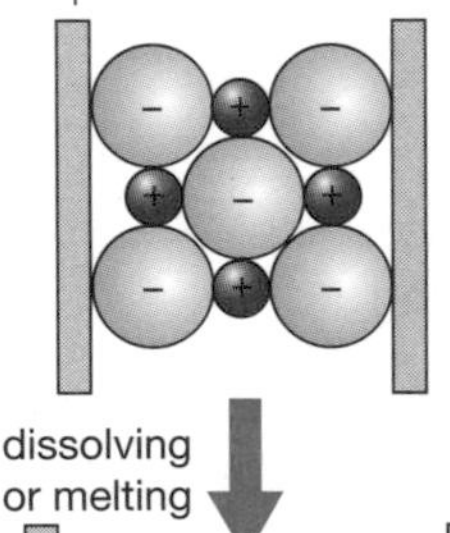

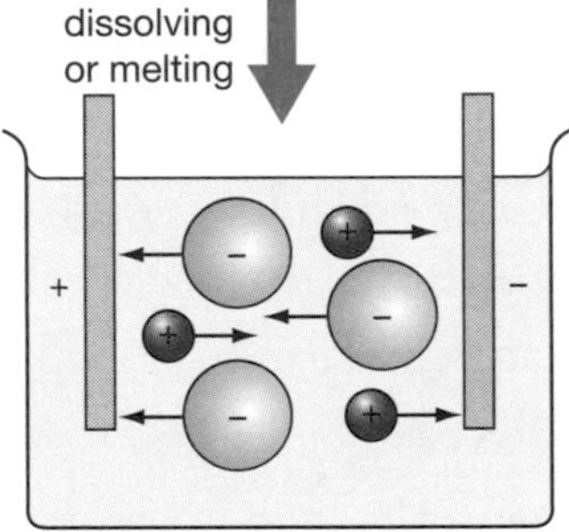

↑ **An ion always moves towards the oppositely charged electrode**

examiner tip

OILRIG:

- **O**xidation **I**s **L**oss of electrons.
- **R**eduction **I**s **G**ain of electrons.

You must include the electrons.

Electrolysing solutions of ions

Revised

- If there is a **mixture of ions** the products formed depend on the **reactivity** of the elements involved.
- If there are a number of metal ions present (they are all positively charged) the element with the **lowest reactivity** forms first.

For example, in a mixture of magnesium chloride and copper chloride it is copper that is produced.

If the mixture is an acidic solution the hydrogen ions are often the least reactive and so hydrogen gas is produced.

examiner tip

Cathode product: hydrogen or non-reactive metal.

Anode product: oxygen or a halogen.

Half-equations

Revised

- Electrode reactions can be represented by **half-equations**.

For example, if a current is passed through hydrochloric acid:

hydrogen is produced at the cathode: $2H^+ + 2e^- \rightarrow H_2$

chlorine is produced at the anode: $2Cl^- \rightarrow Cl_2 + 2e^-$

(If it helps, write anode reactions like this: $2Cl^- - 2e^- \rightarrow Cl_2$)

In the electrolysis of molten lead bromide described above, we have:

cathode: $Pb^{2+}(l) + 2e^- \rightarrow Pb(l)$

anode: $2Br^-(l) \rightarrow Br_2(g) + 2e^-$

Note that the same number of electrons is produced at the anode as are used at the cathode.

You can be asked to complete and balance supplied half-equations, but you will not have to write them from scratch.

Typical half-equations are shown in the table (do not forget the state symbols).

Cathode (–)	Anode (+)
$2H^+(aq) + 2e^- \rightarrow H_2(g)$	$4OH^-(aq) \rightarrow 2H_2O(l) + O_2(g) + 4e^-$
$Cu^{2+}(aq) + 2e^- \rightarrow Cu(s)$	$Cu(s) \rightarrow Cu^{2+}(aq) + 2e^-$
	$2X^-(aq) \rightarrow X_2(g, l \text{ or } s) + 2e^-$ (X is a halogen)

Check your understanding

Tested

56 What is electrolysis? *(1 mark)*

57 What does electrolysis do to compounds? *(1 mark)*

58 Why do electrolytes have to be melted or dissolved in water? *(1 mark)*

59 Copy and complete this table to show the products of electrolysis of different compounds in aqueous solution. *(1 mark per product)*

Mixture	Cathode product	Anode product
Potassium hydroxide (aq)		
Calcium bromide (aq)		
Copper chloride (aq)		
Lead nitrate (aq)		

60 Copy and balance the following half-equations and insert the state symbols.

a) $Na^+ + e^- \rightarrow Na$ *(1 mark)*

b) $Pb^{2+} \rightarrow Pb$ *(3 marks)*

c) $I^- \rightarrow I_2 + e^-$ *(2 marks)*

d) $O^{2-} \rightarrow O_2 + e^-$ *(2 marks)*

e) $OH^- \rightarrow H_2O + O_2$ *(4 marks)*

Answers online **Test yourself online** Online

Electrolysis (2)

Electroplating

Revised

- Electrolysis is used to **electroplate** objects.
- This can be for a variety of reasons and includes **copper** plating and **silver** plating.

For example, to electroplate plate a fork with silver it is made the cathode (negative electrode) in an electrolysis circuit and a piece of silver is made the anode (positive electrode).

As the current flows **silver atoms** from the anode are dissolved to form **silver ions**, $Ag^+(aq)$.

The electrolyte already contains silver ions and as more are made, the same number are changed into silver atoms at the cathode. This coats the fork in a **thin layer** of silver.

anode: $Ag(s) \rightarrow Ag^+(aq) + e^-$

cathode: $Ag^+(aq) + e^- \rightarrow Ag(s)$

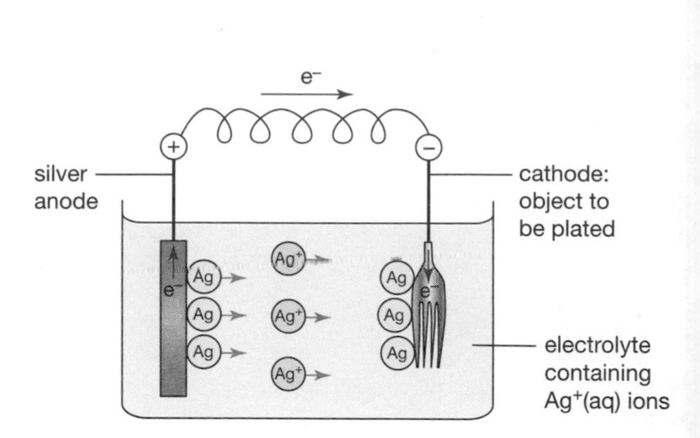

↑ Silver plating a fork. Silver atoms become silver ions at the anode. These positive ions are attracted to the cathode. At the cathode the silver ions become silver atoms. Electrons flow from the anode through the external circuit to the cathode

examiner tip

'Describe and explain' means you have to state a fact and then a reason.

Choose a past question using these command words and look at the mark scheme on the AQA website.

Aluminium manufacture

Revised

- Aluminium is an important metal in today's society.
- Aluminium is manufactured by the **electrolysis** of a molten mixture of **aluminium oxide** and **cryolite**.
- **Aluminium** forms at the **negative** electrode.
- **Oxygen** forms at the **positive** electrode.
- The **positive electrode** (the anode) is made of **carbon**, which reacts with the **oxygen** to produce **carbon dioxide**.

Aluminium oxide is **insoluble** in water and therefore has to be **melted** to make the ions **mobile** (free to move and carry the current). But the **melting point** of aluminium oxide is very high (more than 2000°C) and to melt it would take too much **electrical energy**.

The melting point is brought down to about 950°C by dissolving it in molten **cryolite**, making a large **energy saving**. Cryolite is sodium aluminium fluoride. Using it to dissolve aluminium oxide produces a molten mixture containing both Al^{3+} and Na^+ ions.

Remember that the least reactive element is formed whenever there is a competing mixture of ions in electrolysis. Aluminium is less reactive than sodium and so it is the only metal produced in the process.

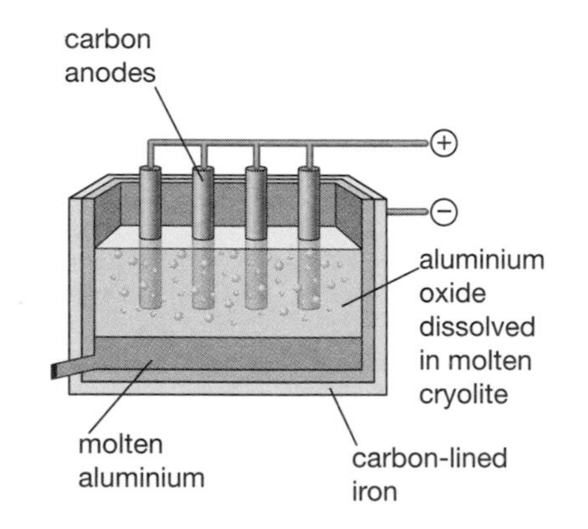

↑ Aluminium oxide is decomposed by electrolysis to produce aluminium

Aluminium extraction electrode reactions

Revised

cathode: $Al^{3+}(l) + 3e^- \rightarrow Al(l)$

anode: $2O^{2-}(l) \rightarrow O_2(g) + 4e^-$

The oxygen reacts with and erodes the anode:

$3C(s) + 3O_2(g) \rightarrow 3CO_2(g)$

The electrons flow from the anode to the cathode through the external circuit.

The electrons used by the cathode have to be balanced by those produced by the anode.

This is achieved by multiplying the half equations:

cathode: $4Al^{3+}(l) + 12e^- \rightarrow 4Al(l)$

anode: $6O^{2-}(l) \rightarrow 3O_2(g) + 12e^-$

Electrolysis of sodium chloride solution

Revised

The electrolysis of **sodium chloride solution**, NaCl(aq) (brine), produces:

- **hydrogen**, $H_2(g)$
- **chlorine**, $Cl_2(g)$
- **sodium hydroxide solution**, NaOH(aq)

These are important reagents for the chemical industry:

- **hydrogen** is used in the manufacture of margarine
- **chlorine** is used in the manufacture of bleach and plastics
- **sodium hydroxide** is used in the manufacture of soap

In the **electrolysis** of **sodium chloride solution**, two elements could be formed at the cathode: sodium (Na) and hydrogen (H_2). But **sodium** is not formed and **hydrogen** is. This is because the **least reactive element** is formed whenever there is a **mixture of ions** in electrolysis.

examiner tip

Make sure you use the phrase 'used in the manufacture of'. There is no sodium hydroxide in the soap you use — it would make your skin very sore.

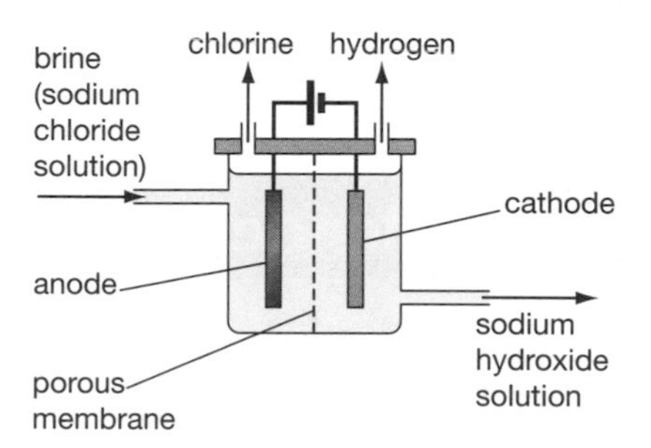

↑ The electrolysis of sodium chloride solution gives three important products

Sodium chloride electrode reactions

anode: $2Cl^-(aq) \rightarrow Cl_2(g) + 2e^-$ chlorine gas is produced

cathode: $2H_2O(l) + 2e^- \rightarrow H_2(g) + 2OH^-(aq)$

At the cathode two substances are produced:

- hydrogen gas
- hydroxide ions (which are released into solution)

This is how the sodium hydroxide is produced.

Check your understanding

Tested

61 What are some advantages of silver-plated cutlery over solid silver cutlery? *(3 marks)*

62 Describe and explain the benefit of adding cryolite to the electrolysis of molten aluminium oxide? *(2 marks)*

63 Why is sodium not produced in the electrolysis of sodium chloride solution? *(1 mark)*

Answers online **Test yourself online** Online

The early periodic table

John Newlands

Revised

John Newlands was an English scientist who discovered the **periodic law**. He listed the then known elements in order of their **atomic weight** and found that every **eighth element** had similar properties — his 'law of octaves'.

The fit between order of atomic weight and the octaves was not perfect — too many **adjustments** had to be made for the elements to fit the octaves law.

His work was **not accepted** by the scientific community. In fact, scientists regarded a periodic table of the elements as a **curiosity**.

Newlands changed the strict order of atomic weights to fit elements into his pattern. Beryllium was called glucium and Di was found not to be an element

Newland's order of atomic weights							
1–7	**8–14**	**15–21**	**22–28**	**29–35**	**36–43**	**42–49**	
H	F	Cl	Co/Ni	Br	Pd	I	Pt/Ir
Li	Na	K	Cu	Rb	Ag	Cs	Tl
G (Be)	Mg	Ca	Zn	Sr	Cd	Ba/V	Pb
Bo	Al	Cr	Y	Ce/La	U	Ta	Th
C	Si	Ti	In	Zr	Sn	W	Hg
N	P	Mn	As	Di/Mo	Sb	Nb	Bi
O	S	Fe	Se	Ro/Ru	Te	Au	Os

Dimitri Mendeleev

Revised

Dimitri Mendeleev was a Russian chemist who published his periodic table in 1869.

He realised that not all the chemical elements had been **discovered** and therefore the pattern of elements was **incomplete**.

He also realised that if the elements were ordered strictly by atomic weight some were put in inappropriate places. For example, iodine is lighter than tellurium, but needs to be placed with the halogens.

Mendeleev left **gaps** for undiscovered elements and **predicted** their discovery. He also accurately **predicted** their position in his periodic table and their **properties**.

↑ Dimitri Mendeleev

I	II	III	IV	V	VI	VII	VIII		
H 1.01									
Li 6.94	Be 9.01	B 10.8	C 12.0	N 14.0	O 16.0	F 19.0			
Na 23.0	Mg 24.3	Al 27.0	Si 28.1	P 31.0	S 32.1	Cl 35.5			
K 39.1	Ca 40.1		Ti 47.9	V 50.9	Cr 52.0	Mn 54.9	Fe 55.9	Co 58.9	Ni 58.7
Cu 63.5	Zn 65.4			As 74.9	Se 79.0	Br 79.9			
Rb 85.5	Sr 87.6	Y 88.9	Zr 91.2	Nb 92.9	Mo 95.9		Ru 101	Rh 103	Pd 106
Ag 108	Cd 112	In 115	Sn 119	Sb 122	Te 128	I 127			
Ce 133	Ba 137	La 139		Ta 181	W 184		Os 194	Ir 192	Pt 195
Au 197	Hg 201	Tl 204	Pb 207	Bi 209					
			Th 232		U 238				

← Mendeleev's periodic table of 1869. Note the gaps and his method of placing the transition elements

Now scientists were beginning to see a periodic table of the elements as a **useful tool** to **predict** the properties of missing elements.

Mendeleev's most famous prediction was about eka-silicon (the element below silicon). Here are some of the predictions he made for eka-silicon compared with the values we know for the element we now call germanium.

	Mendeleev's predictions for eka-silicon (germanium)	Germanium (Ge)
Relative atomic mass	73.4	72.6
Density in g/cm³	5.35	5.32
Melting point in °C	about 800	937
Boiling point in °C	2830	2827

Summary:

- **Newlands**, and then **Mendeleev**, attempted to classify the elements by arranging them in order of their **atomic weights**.
- The list can be arranged in a table so that elements with **similar properties** are in **columns**, known as **groups**.
- The table is called a **periodic** table because similar properties occur at **regular intervals**.
- The early periodic tables were **incomplete** and some elements were placed in **inappropriate** groups if the **strict order** of atomic weights was followed.
- Mendeleev **overcame** some of the problems by **leaving gaps** for elements that he thought had not been discovered.

The modern periodic table

Atomic numbers

Revised

- When electrons, protons and neutrons were discovered early in the **twentieth century**, the periodic table was then **arranged in order** of atomic (**proton**) numbers.
- When this was done, all elements were placed in appropriate groups. (Compare I and Te.)
- The modern periodic table can be seen as an **arrangement** of the elements in terms of their **electronic structures**.
- Elements in the **same group** have the **same number of electrons** in their highest occupied energy level (**outer shell**).
- The number of **electrons** in the highest occupied energy level (**outer shell**) for elements in the main groups is equal to the **group number**.
- Today scientists regard the periodic table (see inside back cover) of the elements as an important summary of the **structure of atoms**.

examiner tip

Elements are arranged in the periodic table by **proton number**. That is the same as the number of electrons **in an atom**. The proton number is constant for an element.

Check your understanding

Tested

1. Why was Newlands' periodic table not accepted by scientists? *(1 mark)*
2. How did Mendeleev order the elements in his periodic table? *(1 mark)*
3. What is the significance of the group number in today's periodic table? *(1 mark)*
4. Explain why the periodic table moved from being 'a curiosity' to 'an important summary of the structure of atoms'. *(3 marks)*

Answers online **Test yourself online** Online

Trends within the periodic table (1)

Group 1: the alkali metals

Revised

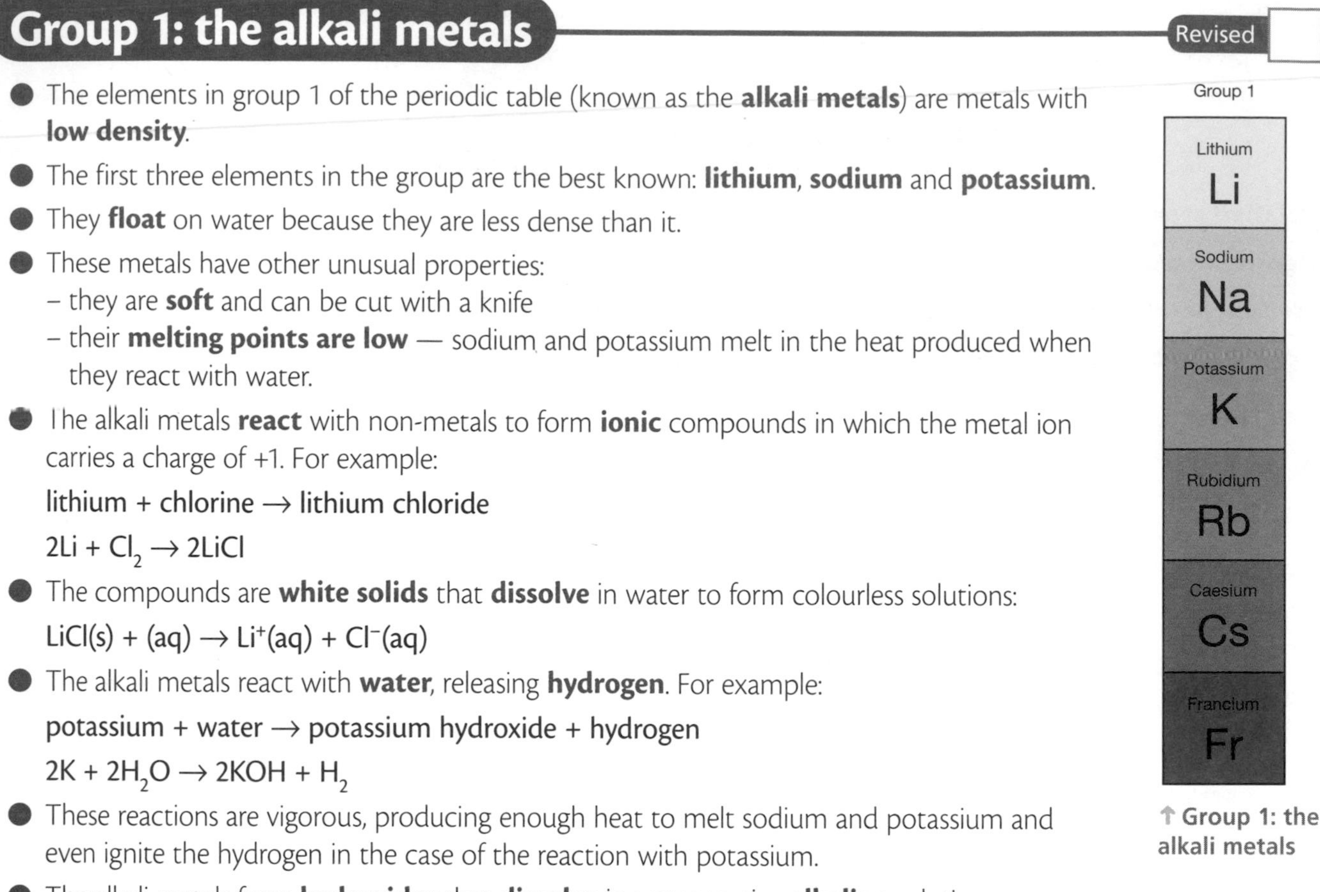

↑ Group 1: the alkali metals

- The elements in group 1 of the periodic table (known as the **alkali metals**) are metals with **low density**.
- The first three elements in the group are the best known: **lithium**, **sodium** and **potassium**.
- They **float** on water because they are less dense than it.
- These metals have other unusual properties:
  - they are **soft** and can be cut with a knife
  - their **melting points are low** — sodium and potassium melt in the heat produced when they react with water.
- The alkali metals **react** with non-metals to form **ionic** compounds in which the metal ion carries a charge of +1. For example:

  lithium + chlorine → lithium chloride

  $2Li + Cl_2 \rightarrow 2LiCl$
- The compounds are **white solids** that **dissolve** in water to form colourless solutions:

  $LiCl(s) + (aq) \rightarrow Li^+(aq) + Cl^-(aq)$
- The alkali metals react with **water**, releasing **hydrogen**. For example:

  potassium + water → potassium hydroxide + hydrogen

  $2K + 2H_2O \rightarrow 2KOH + H_2$
- These reactions are vigorous, producing enough heat to melt sodium and potassium and even ignite the hydrogen in the case of the reaction with potassium.
- The alkali metals form **hydroxides** that **dissolve** in water to give **alkaline** solutions:

  $2KOH(s) + (aq) \rightarrow 2K^+(aq) + 2OH^-(aq)$
- These solutions are **alkaline** because they contain more aqueous **hydroxide** ions, $OH^-(aq)$ than hydrogen ions, $H^+(aq)$. There are no hydrogen ions in the equation.

examiner tip

Acids are **acidic** because of H^+ ions.

Alkalis are **alkaline** because of OH^- ions.

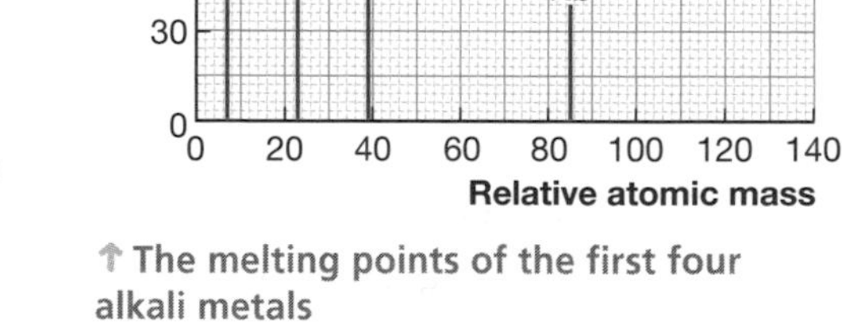

↑ The melting points of the first four alkali metals

The further **down** the group:

- the more **reactive** the element. This is seen with the reactions with air (oxygen): freshly cut pieces of the metals tarnish. Potassium tarnishes **faster** than sodium, which tarnishes **faster** than lithium.
- the **lower** its **melting** point and **boiling** point

Some properties of alkali metals

Alkali metal	Lithium	Sodium	Potassium
Hardness	Cuts with a knife	Softer than Li, cuts easily with a knife	Softer than Na, cuts very easily with a knife
Appearance after cutting	Shiny, dark grey, tarnishes within 30 seconds	Shiny. light grey, tarnishes faster than Li	Shiny, bluish grey, tarnishes faster than Na
Reaction with water	Floats, moves on surface Stays the same shape Fizzes steadily, H_2 Becomes smaller, dissolves No flame Alkaline solution	Floats, moves on surface Melts forming a silvery ball Fizzes vigorously, H_2 Becomes smaller, dissolves No flame Alkaline solution	Floats, moves on surface Melts forming a silvery ball Fizzes violently, H_2 Becomes smaller, dissolves Lilac flame Alkaline solution

The transition elements

Revised

The transition elements are between groups 2 and 3 in the periodic table. They are all metals

Sc 21	Ti 22	V 23	Cr 24	Mn 25	Fe 26	Co 27	Ni 28	Cu 29	Zn 30
Y 39	Zr 40	Nb 41	Mo 42	Tc 43	Ru 44	Rh 45	Pd 46	Ag 47	Cd 48
La 57	Hf 72	Ta 73	W 74	Re 75	Os 76	Ir 77	Pt 78	Au 79	Hg 80

The transition elements have similar properties, and some special properties.

Compared with the elements in Group 1, **transition elements**:

- have **higher melting points** (except for mercury) — for example tungsten (W) melts above 3300°C and is used in traditional light bulbs
- have **higher densities** — for example osmium (Os) and iridium (Ir) are about 22 times denser than water
- are **stronger** and **harder** — for example iron, rather than lithium, is used to build bridges
- are much **less reactive** and so do not react as vigorously with water or oxygen — for example, gold rings do not disappear when you have a shower but iron will rust if not protected

Many transition elements have **ions** with **different** charges, which leads to them:

- forming **coloured compounds**
- being useful as **catalysts**

Some transition metal catalysts

Transition metal	Use as a catalyst
Iron (Fe)	In the Haber process — manufacture of ammonia
Nickel (Ni)	In the hydrogenation of vegetable oils to make margarine
Platinum (Pt) and rhodium (Rh)	In catalytic converters in vehicle exhaust systems

Some transition metal ions and their coloured compounds

Ion	Typical compound	Formula	Colour of compound
Fe^{2+}	Iron(II) hydroxide	$Fe(OH)_2$	Green
Fe^{3+}	Iron(III) hydroxide	$Fe(OH)_3$	Brown
Cu^{+}	Copper(I) oxide	Cu_2O	Red
Cu^{2+}	Copper(II) sulfate	$CuSO_4.5H_2O$	Blue
Mn^{2+}	Manganese(II) sulfate	$MnSO_4.H_2O$	Pink
MnO_4^-	Potassium manganate(VII)	$KMnO_4$	Purple

Check your understanding

Tested

5 For the reaction between potassium and chlorine, write:

a) the word equation *(1 mark)*

b) the balanced chemical equation *(2 marks)*

6 Which is the most reactive metal out of sodium, chromium, lithium and rubidium? *(1 mark)*

7 Which of these metals will float on water: vanadium, zinc, copper, lithium or mercury? *(1 mark)*

8 Why are transition elements rather than alkali metals used for many everyday purposes such as construction? *(3 marks)*

Answers online **Test yourself online** Online

Trends within the periodic table (2)

Group 7: the halogens

Revised

- The second three elements in the group are the best known: **chlorine, bromine** and **iodine**.
- Group 7 elements (**the halogens**) react with metals to form ionic compounds in which the halide ion carries a charge of −1.

An obvious example is:

chlorine + sodium → sodium chloride

$Cl_2 + 2Na \rightarrow 2NaCl$

This would usually be written:

$2Na + Cl_2 \rightarrow 2NaCl$

Another example is:

bromine + iron → iron bromide

$3Br_2 + 2Fe \rightarrow 2FeBr_3$

Again, this would usually be written:

$2Fe + 3Br_2 \rightarrow 2FeBr_3$

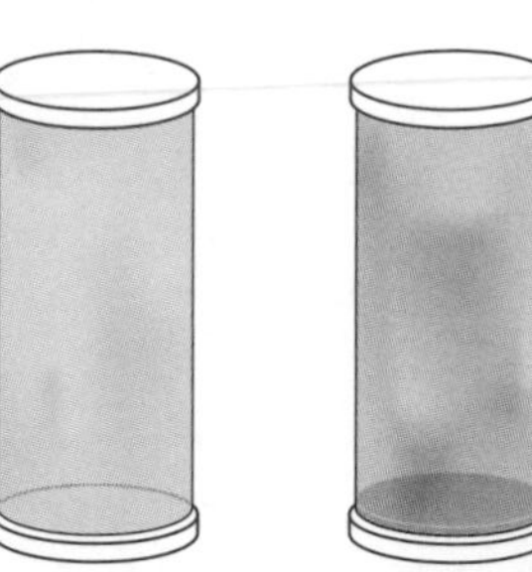
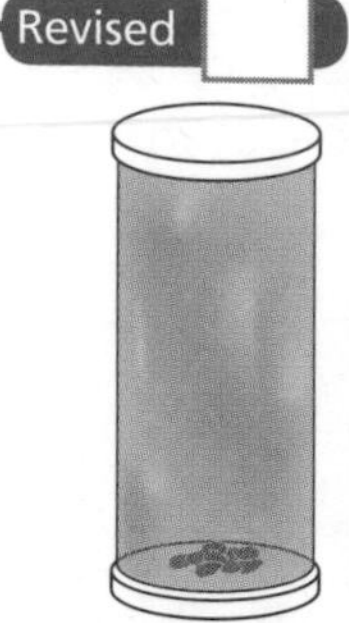

↑ **Halogen vapours. At room temperature chlorine is a gas, bromine is a liquid and iodine is a solid**

examiner tip

The elemental halogens exist as **diatomic molecules**:

- the **formula** for chlorine is Cl_2
- the **symbol** for chlorine is Cl

- The **halogens** are very reactive.

When chlorine reacts with sodium there is a very hot and bright yellow flame.

When chlorine reacts with aluminium foil the metal can melt.

- The further **down** group 7 an element is, the **less reactive** it is.

However, the rate of reaction may increase. This is because the elements increase in density down the group and so the effective concentration is much higher — rate increases with concentration.

Thus, bromine reacts explosively with sodium; bromine reacts vigorously with tin (despite tin being quite unreactive); and iodine will set aluminium alight in the presence of a drop of water.

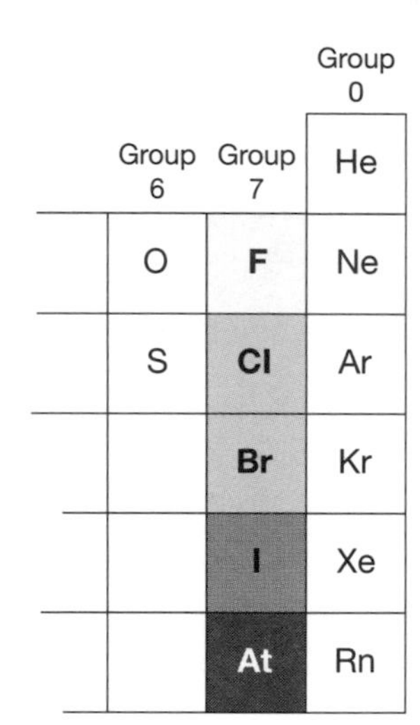

↑ **Group 7: the halogens**

- The further **down** group 7 an element, the **higher** its **melting** point and **boiling** point.

Properties of chlorine, bromine and iodine

Property	Chlorine	Bromine	Iodine
Formula	Cl_2	Br_2	I_2
Relative atomic mass	35.5	80	127
State (at room temperature)	Gas	Liquid	Solid
Colour at room temperature	Pale green	Dark red	Dark grey
Colour of vapour	Pale green	Orange	Purple
Melting point in °C	−101	−7	114
Boiling point in °C	−35	59	184
Density in g/cm^3	1.57 (as liquid at boiling point)	3.14	4.94
Odour	Strong 'swimming pool'	Exceedingly strong 'swimming pool'	Faintly antiseptic

Halogen displacement reactions

Revised

- A more reactive halogen can displace a less reactive halogen from an aqueous solution of its salt.

This is the same as any other displacement reaction — the most reactive element ends up in a compound.

For example, chlorine displaces bromine from a solution of potassium bromide:

chlorine	+	potassium bromide	→	potassium chloride	+	bromine
$Cl_2(g)$	+	$2KBr(aq)$	→	$2KCl(aq)$	+	$Br_2(aq)$
pale green		colourless		colourless		brown solution

In the same way bromine displaces iodine from a solution of sodium iodide:

bromine	+	sodium iodide	→	sodium bromide	+	iodine
$Br_2(g)$	+	$2NaI(aq)$	→	$2NaBr(aq)$	+	$I_2(aq)$
brown		colourless		colourless		almost black solid

Groups 1 and 7 reactivity trends

Revised

- Down **group 1** reactivity **increases.**
- Down **group 7** reactivity **decreases**.
- Atoms can react by **gaining** or **losing** one or more **outer electrons**.
- Going down both groups 1 and 7 the atoms **increase** in size — the outer electrons are further from the nucleus and there are more inner electrons shielding them from the attraction of the positive charge.
- **Group 1** elements react by **losing** their single outer electron.
- As this electron becomes **further** from the positive nucleus it is in a higher **energy level.**
- This electron is **less strongly** attracted to the atom. It is **more easily lost**.
- **Group 7** elements react by **gaining** an outer electron.
- Going down the group this electron enters a higher **energy level.**
- This electron is **less strongly** attracted to the atom. It is **less easily gained**.

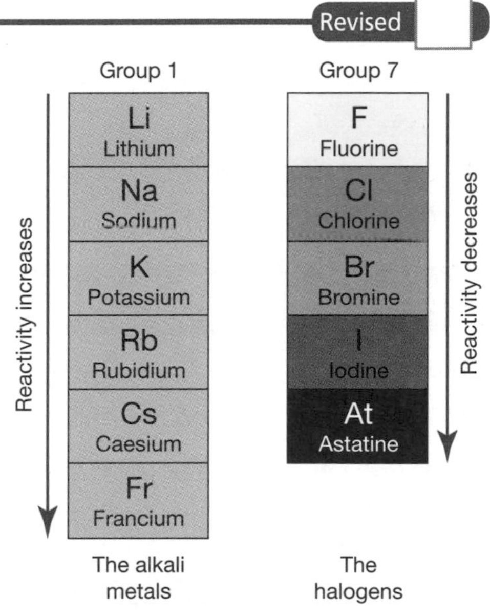

↑ **Group 7: the halogens**

examiner tip

Always remember group trends **down** the group.

Check your understanding

Tested

9 For the reaction between copper and fluorine, write:

a) the word equation *(1 mark)*

b) the balanced chemical equation *(2 marks)*

10 What is the formula of iodine? *(1 mark)*

11 What is the formula of a bromide ion? *(1 mark)*

12 For the reaction between fluorine gas and rubidium iodide in solution, write:

a) the word equation *(1 mark)*

b) the balanced chemical equation *(2 marks)*

13 Explain why caesium is more reactive than sodium. *(2 marks)*

14 Explain why iodine is less reactive than chlorine. *(2 marks)*

Answers online **Test yourself online** Online

Hard and soft water

- **Hard water** contains **dissolved** compounds, usually of **calcium** or **magnesium**.
- The compounds are dissolved when water comes into contact with rocks.

Rainwater is naturally slightly acidic and so it can dissolve minerals that are in rocks to make a solution containing ions such as calcium Ca^{2+}, magnesium Mg^{2+}, hydrogencarbonate HCO_3^- and sulfate SO_4^{2-}.

- **Hard water** reacts with soap to form **scum** and so more soap is needed to form lather.

Soap is the sodium salt of long-chain organic acids, for example, sodium stearate, $C_{17}H_{35}COONa$.
Scum is formed by **precipitation** of insoluble **metal salts** of these acids:

stearate ions + calcium ions → calcium stearate (scum)

$$2C_{17}H_{35}COO^-(aq) + Ca^{2+}(aq) \rightarrow (C_{17}H_{35}COO)_2Ca(s)$$

A lather forms only after the soap has reacted with the metal ions in the water.

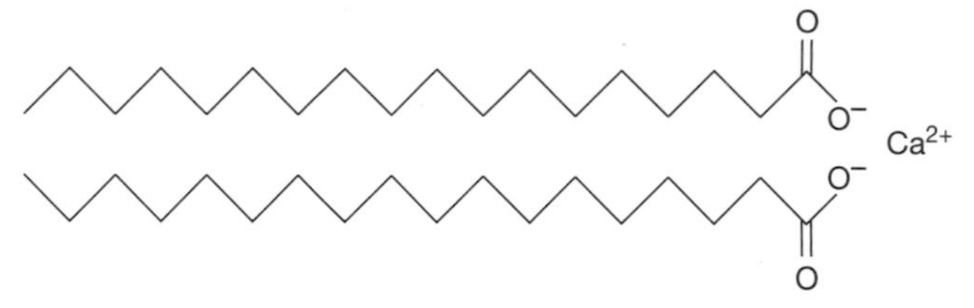

- **Soft water** readily forms a lather with soap — there are **few dissolved** ions to react with the soap.
- **Soapless detergents** do not form scum — there are **no stearate** ions to react with dissolved metal ions.

Permanent and temporary hardness

Revised

- There are two types of hard water — **permanent** and **temporary.**
- **Permanent** hard water remains hard when it is boiled.
- **Temporary** hard water is softened by boiling.

Permanent hard water is caused by rainwater dissolving the sparingly soluble rock gypsum, which is calcium sulfate:

$$CaSO_4(s) + aq \rightarrow Ca^{2+}(aq) + SO_4^{2-}(aq)$$

Temporary hard water is formed by reaction between carbon dioxide dissolved in rainwater and calcium carbonate, giving calcium hydrogencarbonate in solution:

$$CO_2(aq) + H_2O(l) + CaCO_3(s) \rightarrow Ca(HCO_3)_2(aq)$$

- Temporary hard water contains hydrogencarbonate ions (HCO_3^-)
- These ions decompose on heating to produce carbonate ions, which react with calcium and magnesium ions to form precipitates:

  $HCO_3^-(aq) \rightarrow H^+(aq) + CO_3^{2-}(aq)$

  then:

  $CO_3^{2-}(aq) + Ca^{2+}(aq) \rightarrow CaCO_3(s)$

  and

  $CO_3^{2-}(aq) + Mg^{2+}(aq) \rightarrow MgCO_3(s)$

Benefits and drawbacks of using hard water

Revised

- Using hard water can **increase costs** because more soap is needed.

This may not be significant at home, but on an industrial scale it certainly is.

- When temporary hard water is heated it can produce **scale** that **reduces the efficiency** of heating systems and kettles.

Scale (limescale) builds up on the insides of heating systems and kettles. It reduces the amount of water that can flow through the pipes.

Every time a kettle is heated the scale has to be heated as well — using unnecessary energy.

- Hard water has some **benefits** because calcium compounds are good for the development and maintenance of **bones** and **teeth** and also help to **reduce heart disease**.

Water softening

Revised

Hard water is **softened** by **removing** the dissolved calcium and magnesium ions. **Temporary hard** water can be softened simply by **boiling** it (see above). **Both** types can be softened by:

- adding **sodium carbonate**, which reacts with the calcium and magnesium ions to form a precipitate of calcium carbonate and magnesium carbonate

sodium carbonate + calcium ions → calcium carbonate + sodium ions

$Na_2CO_3(aq) + Ca^{2+}(aq) \rightarrow CaCO_3(s) + 2Na^+(aq)$

sodium carbonate + magnesium ions → magnesium carbonate + sodium ions

$Na_2CO_3(aq) + Mg^{2+}(aq) \rightarrow MgCO_3(s) + 2Na^+(aq)$

- using commercial water softeners such as **ion-exchange** columns containing hydrogen ions or sodium ions, which replace the calcium and magnesium ions when hard water passes through the column

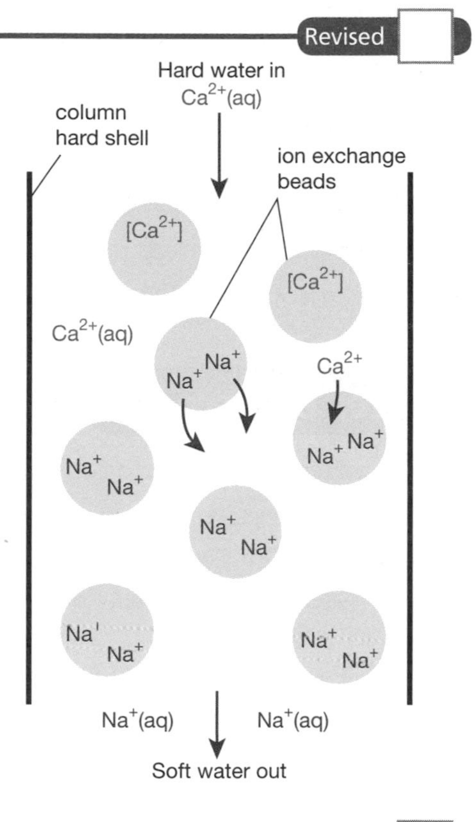

→ An ion-exchange column

Evaluating hard water

Revised

Advantages of hard water	Disadvantages of hard water
Good for teeth and bones	Wastes soap
Less likely to dissolve lead pipes (in old houses)	Causes kettles to 'fur up' and blocks pipes
Reduces risk of heart disease	Scum is left in baths and showers
Better beer can be brewed	Clothes may feel hard after washing
	Softening requires capital outlay and running costs
	Water softened using a sodium ion-exchange column is not safe for babies
	Requires more energy to get things clean

examiner tip

Another topic for evaluation. Remember: consider the pros and cons using the information in the question. A table really is a good idea.

Check your understanding

Tested

15 What is hard water? *(2 marks)*

16 How is scum formed? *(2 marks)*

17 Which type of hard water can be softened by boiling? *(1 mark)*

18 Explain how boiling hard water softens it. *(5 marks)*

19 What are some advantages and disadvantages of living in a hard water area? *(4 marks)*

Answers online **Test yourself online** Online

Purifying water

Water for life

Revised

- Water of the correct **quality** is essential for life.

Our bodies are about 60% by weight water. All the reactions in our bodies happen in aqueous solution, our blood is mainly water and even the insides of our lungs are wet.

- For humans, drinking water should have sufficiently **low levels** of **dissolved salts** and **microbes**.

We can safely drink hard water but too much mineral content in water can lead to health problems.

Microbes in water can lead to diseases such as **typhoid**, **cholera** and **dysentery**.

In the UK the water companies supply water to our taps that is safe to drink.

Water treatment

Revised

- The water we drink has to come from **appropriate** supplies.

Water companies take water from **rivers** and **reservoirs**. This ensures that the pre-treated water does not have too many dissolved salts or microbes.

- The water is put through a number of processes to remove various types of contaminant.

It is passed through various **filter beds** to remove solids. A coagulant is then added to make remaining impurities clump together. The coagulated materials are then removed.

- **Chlorine** may be added to drinking water to reduce the number of microbes — it kills them.

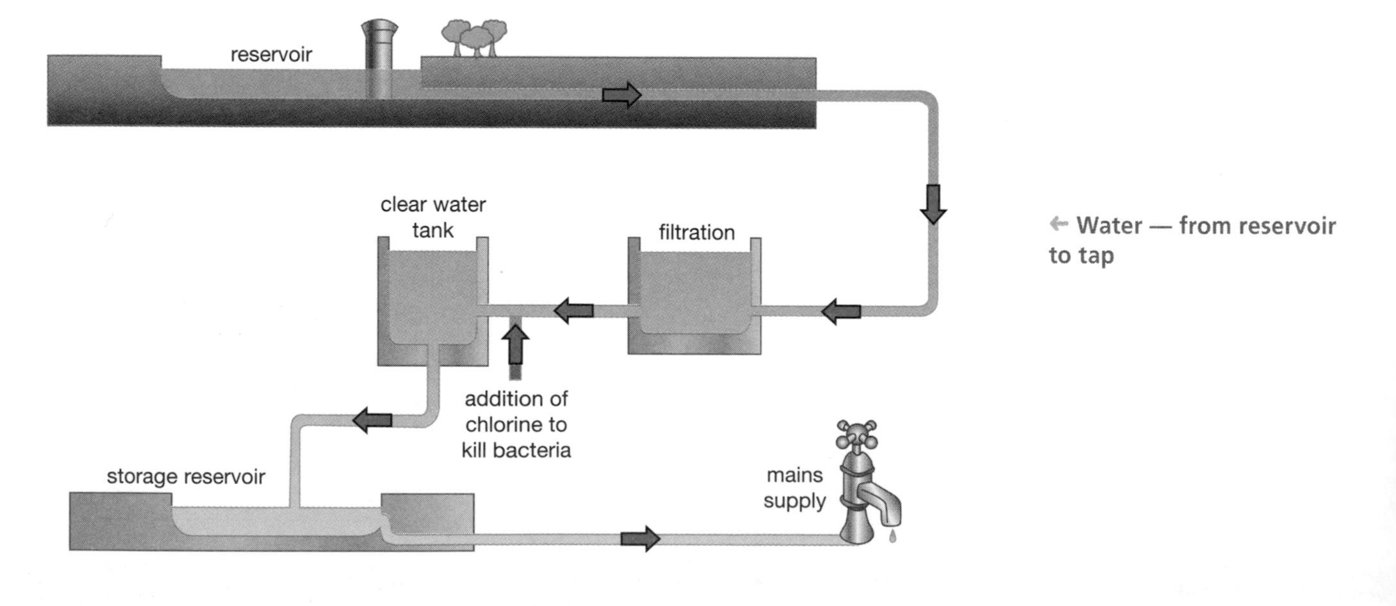

← Water — from reservoir to tap

Fluoride in drinking water

Revised

- Fluoride may be added to the water supply to **improve dental health**.

The fluoride ion (F^-) can be added to the chemical structure of tooth enamel. It adds to the apatite to form fluorapatite ($Ca_5(PO_4)_3F$). This increase in the amount of fluoride in teeth makes them more **resistant to decay** (acid attack).

While children are growing, their teeth can absorb fluoride from food and water.

However, the addition of fluoride to drinking water is a **contentious** method of improving dental health. Some people say it can lead to there being **too much fluoride** in the water — this can cause **fluorosis**, or spots on children's teeth, and can lead to **bone abnormalities** and **brittleness**.

Another argument against the fluoridation of water supplies is based on **ethics**. The reason given is that people are having medication without having **chosen** it. Other people may want to keep themselves **free** from **unnatural** chemicals.

Water filters

Revised

- Tap water in the UK is perfectly safe to drink but some people prefer to filter their tap water before they use it.
- Water filters containing **carbon**, **silver** and **ion-exchange resins** can remove some **dissolved substances** from tap water to improve the taste and quality.

The carbon is in the form of **activated charcoal,** which has a very high surface area. It removes **chlorine** and **organic** chemicals.

The **silver** is used on the charcoal to prevent the growth of **bacteria** that are trapped by the carbon.

Ion-exchange resins soften the water — preventing unsightly, but harmless, scum on hot drinks.

examiner tip

Examiners like to write questions that cross topic boundaries. Refer back to softening water and look at ion-exchange resins.

Distillation

Revised

- **Pure water** can be produced by **distillation**.

This has been used in laboratories to make up aqueous solutions — ensuring they do not contain unwanted ions. The process involves boiling impure water and then condensing the vapour.

Distilled water is **expensive** — the **energy** costs are high and a lot of **cooling water** is used.

- Deionised water has largely replaced distilled water in laboratories. This is made by passing tap water through a special ion-exchange resin to remove the ions. It is a fast and inexpensive method of producing reasonably pure water.

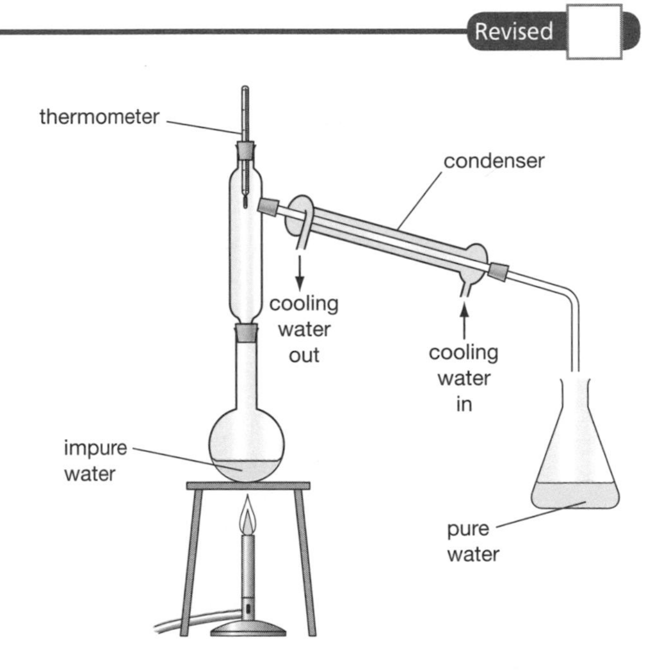

→ **Distillation to produce pure water requires a lot of energy and water**

Check your understanding

Tested

20 Why do water companies not use industrial waste-water or take water from sewers? *(2 marks)*

21 How does fluoride in water benefit public health? *(1 mark)*

22 How do water filters improve the taste and quality of water? *(4 marks)*

23 Why is distilled water expensive? *(1 mark)*

Answers online **Test yourself online** Online

Energy from reactions

Energy of combustion

Revised

- Energy is normally measured in **joules** (J).

The symbol is a capital letter J because the unit is named after a person, the English scientist James Joule.

- The relative amounts of **energy released** when substances burn can be measured by simple **calorimetry**, for example by heating water in a glass or metal container.
- This method can be used to compare the amount of energy released by **fuels** and **foods**.

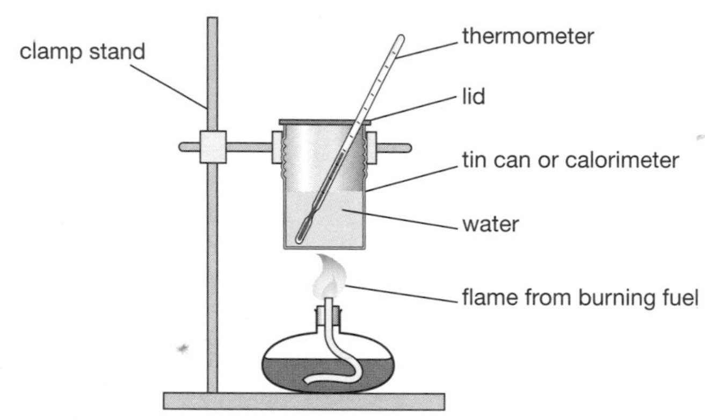

← **Measuring energy released when substances burn**

This method uses the water to 'catch' the energy that has been released by the burning material. It assumes that **all** the energy released by the fuel or food has been **absorbed** by the water. This energy **heats** the water.

You know that heating a full kettle of water takes more energy than heating one half full. You also know that boiling a kettle of water takes more energy than just heating it a little.

These factors are combined in a simple equation that lets us calculate the amount of energy (**Q**) used by the kettle, or absorbed in our **calorimetry** experiment:

$$Q = mc\Delta T$$

where Q = **energy** (in kJ) absorbed by the water (= energy released by the food or fuel)

m = **mass** (in kg) of water

c = **specific heat capacity** of water (= 4.2 kJ/kg/°C)

ΔT = **change in temperature** of the water (in °C)

Example

When 5.0 g of a camping fuel gel was burned, the heat produced raised the temperature of 200 g (0.2 kg) of water from 29°C to 59°C. The gel contained 1.5 g of ethanol.

Calculate: **(i)** the energy released by the gel; **(ii)** the energy released by 1 mole of ethanol (M_r = 46 g/mol).

(i) energy released $= mc\Delta T$

$= 0.2 \times 4.2 \times 30$

$= 25.2$ kJ

examiner tip

energy released by the fuel = energy absorbed by the water

(ii) energy released per mole of ethanol = (energy released/mass of fuel burned) × M_r

$$= \frac{25.2\text{ kJ}}{1.5\text{ g}} \times 46\text{ g/mol}$$

$= 773$ kJ/mol

Energy of reaction

Revised

- The amount of **energy** produced by a **chemical reaction in solution** can be calculated from the measured temperature change of the solution when the reagents are mixed in an **insulated** container.
- This method can be used for reactions of **solids with water** or for **neutralisation** reactions.

This method also uses the water in the solution to 'catch' the energy that has been released by the reaction. It, again, assumes that **all** the energy released by the reaction has been **absorbed** by the water and that none has heated the container or surrounding air.

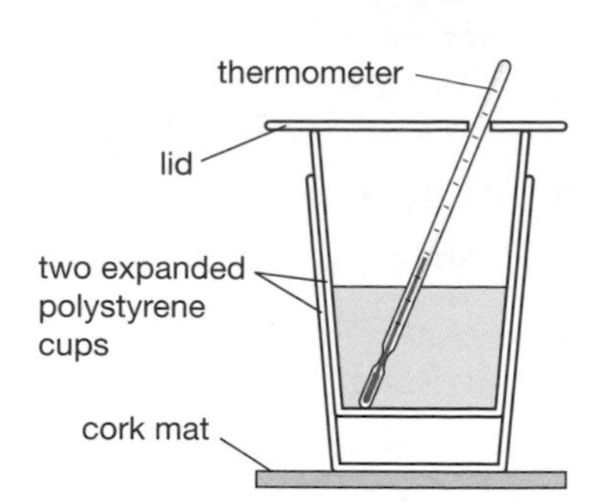

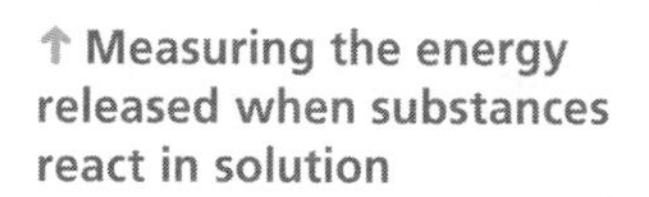
Measuring the energy released when substances react in solution

Example

When 4.5 g of lithium chloride (LiCl, M_r = 42.5 g/mol) was dissolved in 50 g of water the temperature increased by 13°C.

Calculate: **(i)** the energy released by the reaction; **(ii)** the energy released by dissolving 1 mole of LiCl.

(i) energy released $= mc\Delta T$

$= 0.05 \times 4.2 \times 13$

$= 2.73\,\text{kJ}$

(ii) energy released per mole of LiCl $= \dfrac{\text{energy released}}{\text{mass of LiCl}} \times M_r$

$= \dfrac{2.73\,\text{kJ}}{4.5\,\text{g}} \times 42.5\,\text{g/mol}$

$= 25.8\,\text{kJ/mol}$

examiner tip

Always include the units in calculations.

Check that the final units are what you expect.

Here (kJ/g) × (g/mol) = kJ/mol.

Energy level diagrams

Revised

- Simple energy level diagrams can be used to show the **relative energies** of reactants and products, the **activation energy** and the overall **energy change** of a reaction.
- In an **exothermic** reaction energy is **given out** to the surroundings.
- The reactants start at a **higher** energy level than where the products finish.
- In an **endothermic** reaction energy is **absorbed** from the surroundings.
- The reactants start at a **lower** energy level than where the products finish.

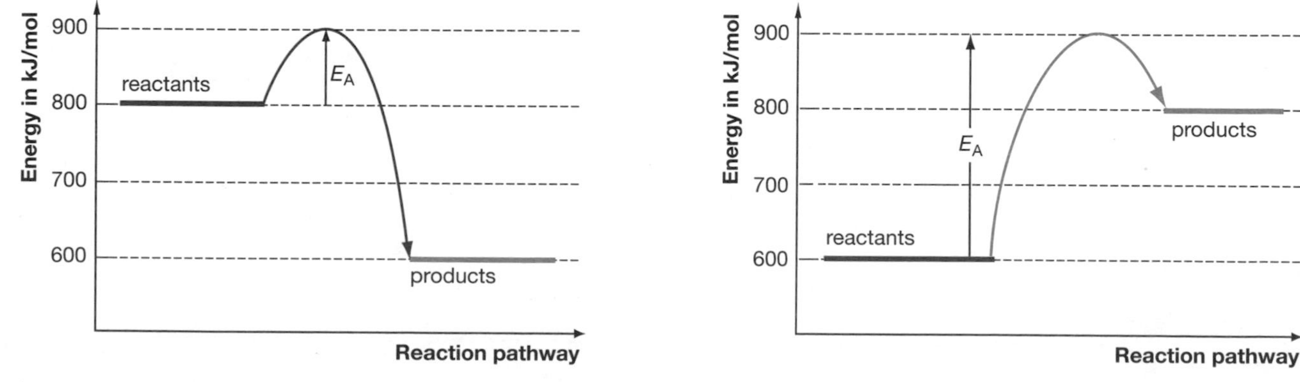

An energy level diagram for an exothermic reaction

An energy level diagram for an endothermic reaction

In these diagrams E_A is the **activation energy**, the energy required to start the reaction.

Breaking and making bonds

Revised

- During a chemical reaction:
  – energy must be **supplied** to **break bonds**
  – energy is **released** when **bonds** are **formed**
- The **difference** between these energies is the **overall energy change**.
- In an **exothermic** reaction, the energy **released** from **forming** new bonds is **greater** than the energy **needed** to **break** existing bonds.
- In an **endothermic** reaction, the energy **needed** to **break** existing bonds is **greater** than the energy **released** from **forming** new bonds.

Example

Methane burns in air (oxygen) to release (heat) energy.

The energy level diagram below shows that energy is absorbed when bonds are broken and released when new bonds are made. It also shows that more energy is released than absorbed so, overall, heat is produced.

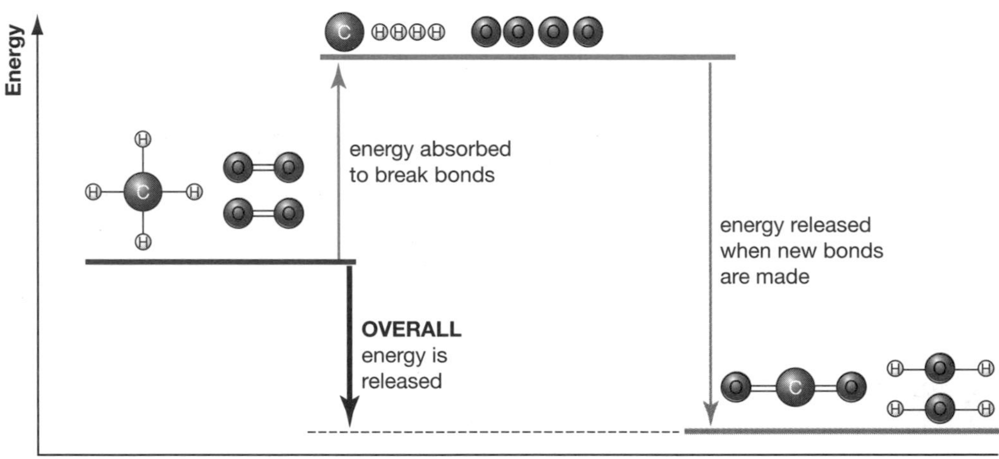

Energy level diagram for methane burning — with bonds being broken and new bonds made

The bond energies are listed in the table on the right.

Bond energy is the amount of energy associated with 1 mole of the bonds.

Bond energy is either absorbed to break one mole of bonds, or released when one mole of the bonds is made.

The calculation can be tabulated, as shown below:

Bond	Bond energy in kJ/mol
C–H	412
O=O	496
O–H	463
C=O	743

Breaking bonds			Making bonds		
Bonds	Energy in kJ		Bonds	Energy in kJ	
4 × C–H	4 × 412	+1648	2 × C=O	2 × –743	–1486
2 × O=O	2 × 496	+992	4 × O–H	4 × –463	–1852
	Total	+2640		Total	–3338
Overall energy change = 2640 – 3338 = –698 kJ (exothermic)					

examiner tip

Breaking bonds requires energy. It is an endothermic process.

Making bonds gives out energy. It is an exothermic process.

Catalysts

Revised

- **Catalysts** provide a **different pathway** for a chemical reaction that has a **lower activation energy**.
- Catalysts speed up a reaction because, without increasing the temperature, more collisions reach the **activation energy**.

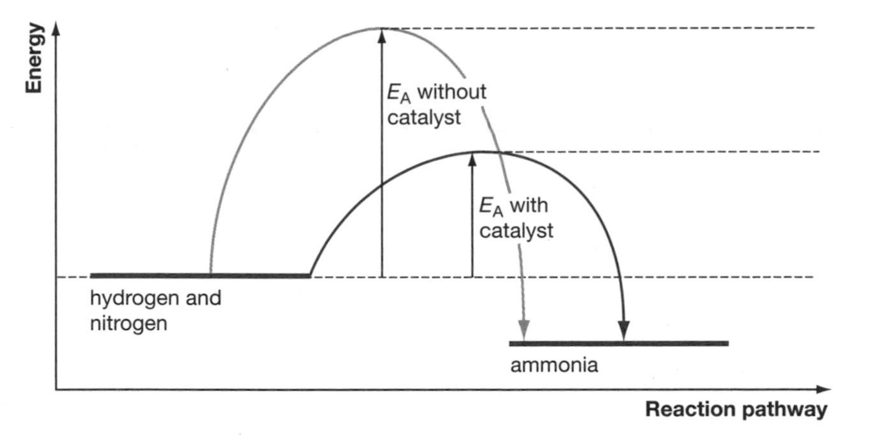

The catalyst lowers the activation energy, so more collisions result in a reaction. The rate of reaction increases

examiner tip

A catalyst **is a substance** that increases rate but is not used up.

A catalyst **works** by lowering activation energy.

Hydrogen as a fuel

Revised

- Hydrogen can be burned as a **fuel** in combustion engines:

  hydrogen + oxygen → water
- It can also be used in **fuel cells** that produce **electricity** to power vehicles.

The use of hydrogen as a fuel appears to be **problem free** — water is the only product, along with the energy released. But the questions of how to **make** the hydrogen and where to **store** it have to be answered.

Most hydrogen comes from reacting methane (CH_4) with steam at 1000°C. That takes a lot of energy and consumes a non-replaceable fossil fuel.

Storing hydrogen requires high-pressure steel tanks, but research is under way into absorbing it in special materials.

Check your understanding

Tested

24 Two aqueous solutions (75 g of each) were reacted together in a perfectly insulated beaker. The temperature decreased by 12°C. How much energy was absorbed by the reaction? *(2 marks)*

25 150 g of water were heated by burning 3.0 g of propan-1-ol (C_3H_7OH). The initial temperature if the water was 18.5°C, the final temperature was 68.5°C. How much energy was released by the propan-1-ol and how much energy would be released by burning 1 mole of it? *(6 marks)*

26 Calculate overall energy change for the reaction:

$C_2H_5OH + 3O_2 \rightarrow 2CO_2 + 3H_2O$

Use a table to work out your answer. *(10 marks)*

Bond	Bond energy in kJ/mol
C–C	348
C–H	412
C–O	360
O=O	496
O–H	463
C=O	743

27 How do catalysts work? *(1 mark)*

Answers online **Test yourself online** Online

Analysing substances (1)

Flame tests

Revised

- Flame tests can be used to identify **metal** ions.

Flame tests are very **sensitive**, i.e. they detect very **small** amounts of the metal ion in minute samples. Therefore they are useful in **forensic** analysis.

Flame tests use an inert wire (platinum is good):

Step 1 Dip the wire into hydrochloric acid.

Step 2 Heat the wire in a roaring Bunsen flame.

Step 3 Dip the wire back in the acid.

Step 4 Dip the wire into the test sample.

Step 5 Hold the wire in the edge of the blue Bunsen flame and observe the flame colour.

The distinctive coloured flames produced by some metal ions

Metal ion	Lithium (Li^+)	Sodium (Na^+)	Potassium (K^+)	Calcium (Ca^{2+})	Barium (Ba^{2+})
Flame colour	Crimson	Yellow	Lilac	Red	Green

More **ions** can be identified by reacting them with carefully chosen chemicals and observing the reaction.

Testing with sodium hydroxide

Revised

- Most metal ions give precipitates with **sodium hydroxide** solution.
- Some of the precipitates **dissolve** in **excess** sodium hydroxide.

examiner tip

Describing solutions:

- Water is **clear** and **colourless**.
- copper sulfate solution is **clear** and **blue**.

Metal ions react with sodium hydroxide solution to form insoluble hydroxides

Solution	Appearance of solution	Observations on adding a few drops of sodium hydroxide	Observations on adding excess sodium hydroxide	Name of precipitate
Al^{3+}	Clear colourless	White precipitate $Al^{3+} + 3OH^- \rightarrow Al(OH)_3$	Precipitate dissolves to clear colourless solution	Aluminium hydroxide
Ca^{2+}	Clear colourless	White precipitate $Ca^{2+} + 2OH^- \rightarrow Ca(OH)_2$	Precipitate remains	Calcium hydroxide
Mg^{2+}	Clear colourless	White precipitate $Mg^{2+} + 2OH^- \rightarrow Mg(OH)_2$	Precipitate remains	Magnesium hydroxide
Cu^{2+}	Clear blue	Blue precipitate $Cu^{2+} + 2OH^- \rightarrow Cu(OH)_2$	Precipitate remains	Copper(II) hydroxide
Fe^{2+}	Clear pale green	Green gelatinous precipitate $Fe^{2+} + 2OH^- \rightarrow Fe(OH)_2$	Precipitate remains	Iron(II) hydroxide
Fe^{3+}	Clear orange	Brown gelatinous precipitate $Fe^{3+} + 3OH^- \rightarrow Fe(OH)_3$	Precipitate remains	Iron(III) hydroxide

Testing for carbonates

Revised

When an acid reacts with a carbonate, carbon dioxide is produced. This has to be identified using limewater.

- **Carbonates** react with **dilute acids** to form **carbon dioxide**:

  $MCO_3 + 2HCl \rightarrow MCl_2 + CO_2 + H_2O$
  where M is a metal

- Carbon dioxide produces a white precipitate with **limewater**:

  $CO_2 + Ca(OH)_2 \rightarrow CaCO_3 + H_2O$

- This turns limewater **cloudy**. The cloudiness is very small particles of calcium carbonate.

↑ Testing for carbonate using a dilute acid and limewater

Testing for halides

Revised

- **Halide ions** in solution produce **precipitates** with **silver nitrate** solution in the presence of dilute **nitric acid**.

Test	Chloride (Cl^-)	Bromide (Br^-)	Iodide (I^-)
Nitric acid plus a few drops of silver nitrate solution	White precipitate of silver chloride	Cream precipitate of silver bromide	Pale yellow precipitate of silver iodide

The general chemical reaction for these tests is:

silver ion + halide ion → silver halide

$Ag^+(aq) + X^-(aq) \rightarrow AgX(s)$ where X is a halide ion

Testing for sulfates

Revised

- **Sulfate** ions in solution produce a **white precipitate** with **barium chloride** solution in the presence of dilute **hydrochloric acid**.

  barium chloride + sulfate ion → barium sulfate

  $Ba^{2+}Cl_2(aq) + SO_4^{2-}(aq) \rightarrow BaSO_4(s)$

examiner tip

Sulfuric acid (H_2SO_4) is a sulfate and gives a white precipitate with barium ions.

Check your understanding

Tested

28 What tests would you carry out, and in what order would you do them, to distinguish between three white solids: aluminium nitrate, calcium nitrate and magnesium nitrate? *(4 marks)*

29 Some bags of fertiliser (potassium nitrate) and weed killer (sodium chlorate(v)) have lost their markings. What test would identify the contents of each bag and what would be seen during the test? *(3 marks)*

30 For the reaction of silver nitrate with a solution of potassium bromide, write:

a) the word equation *(1 mark)*

b) the balanced chemical equation *(2 marks)*

Answers online **Test yourself online** Online

Analysing substances (2)

Volumetric analysis (titrations)

Revised

- When an acid reacts with an alkali, a salt and water are produced.

For example:

nitric acid + potassium hydroxide → potassium nitrate + water

$HNO_3(aq)$ + $KOH(aq)$ → $KNO_3(aq)$ + $H_2O(l)$

This is a **neutralisation** reaction — the resulting solution is neutral, having a **pH of 7**.

pH 7 is achieved only if **exactly** the same amount (in moles of course) of nitric acid and potassium hydroxide are reacted.

If there is just a small amount of either in excess, then the final pH will be above or below 7 — the solution will not be neutral.

- An indicator changes colours in an acid or an alkali.
- Litmus is red in acids and dark blue in alkalis.

The volumes of acid and alkali solutions that react with each other can be measured by titration using a suitable indicator.

examiner tip

Remember to use the correct words.

- Acids are **acidic**.
- Alkalis are **alkaline**.

Not all indicators are suitable: a good indicator has one, sharp, easily seen colour change at the right pH.

Universal indicator will not work — it has a gradually changing range of colours over the range of the pH scale.

pH 0 1 2 3 4 5 6 7 8 9 10 11 12 13 14

strong acid | weak acid | neutral | weak alkali | strong alkali

increasing acidity ← → increasing alkalinity

0 1 2 3 4 5 6 7 8 9 10 11 12 13 14

universal indicator

↑ **Universal indicator changes colour gradually over the pH range**

Some indicators and their colours in acidic and alkaline conditions

Indicator	Colour in acid	Colour in alkali	Colour change
Universal indicator	Red to light green	Dark green to blue	Gradual
Litmus	Red	Blue	Unclear
Methyl orange	Red	Yellow	Sharp
Phenolphthalein	Colourless	Pink	Sharp
Bromothymol blue	Yellow	Bright blue	Sharp

- Titrations are used to accurately find the **amount** (always in **moles**) of a chemical in a certain volume.
- If the concentration of one of the reactants is known, the results of a titration can be used to find the concentration of the other reactant.

Titration calculations

Revised

- These really are straightforward and there are only three or four steps.
- All you have to know is:

amount = concentration × volume

mol = (mol/dm^3) × dm^3

examiner tip

Amount is only ever in **moles**.

Volume is in cm^3 but calculated using dm^3.

Mass is in grams.

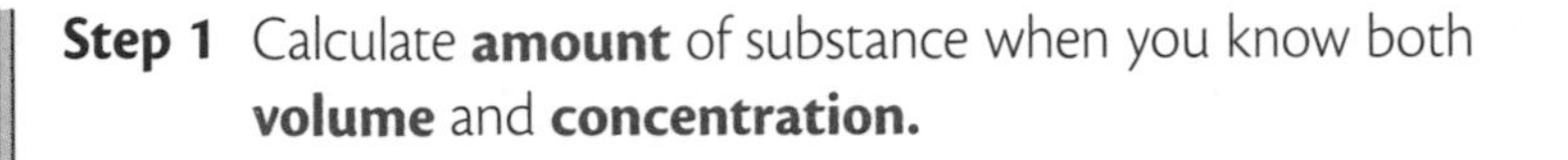

Step 1 Calculate **amount** of substance when you know both **volume** and **concentration.**

Step 2 Use the **balanced chemical equation** to work out the **amount** of the other substance in the reaction.

Step 3 Calculate the **concentration** (mol/dm^3) of this second substance, since you know both its amount and volume.

Step 4 You may have to calculate this concentration as a **mass concentration** (g/dm^3). Just use M_r to change the amount into mass:

$$\text{mass concentration} = M_r \times \text{concentration}$$
$$\text{g/dm}^3 = (\text{g/mol}) \times (\text{mol/dm}^3)$$

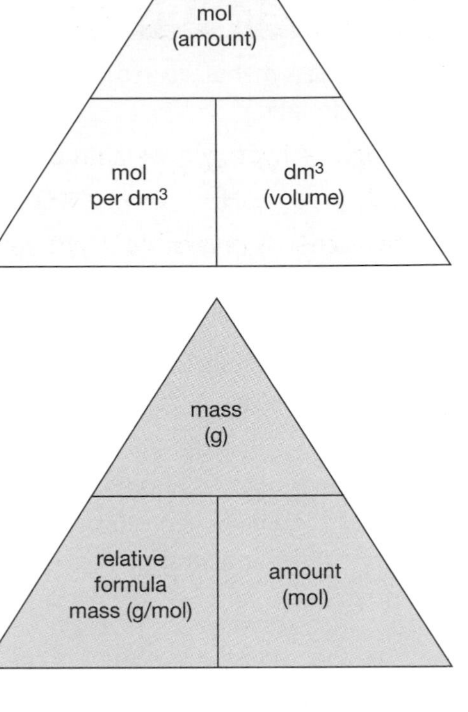

Example

25.0 cm^3 of 0.101 mol/dm^3 NaOH were exactly neutralised using 15.2 cm^3 H_2SO_4.

Calculate the concentration of the sulfuric acid in mol/dm^3 and then in g/dm^3.

$$H_2SO_4 + 2NaOH \rightarrow Na_2SO_4 + 2H_2O$$

Step 1 amount of NaOH = volume × concentration

$$= 25.0\,\text{cm}^3 \times \frac{0.101\,\text{mol/dm}^3}{1000\,\text{cm}^3\text{ per dm}^3}$$
$$= 2.53 \times 10^{-3}\,\text{mol}$$

Step 2 From the equation, the amount of H_2SO_4 = ½ the amount of NaOH

$$= \frac{2.53 \times 10^{-3}\,\text{mol}}{2}$$
$$= 1.27 \times 10^{-3}\,\text{mol}$$

Step 3 Concentration (mol/dm^3) of this H_2SO_4 $= \dfrac{\text{amount}}{\text{volume}}$

$$= \frac{1.27 \times 10^{-3}\,\text{mol}}{15.2\,\text{cm}^3} \times 1000\,\text{cm}^3/\text{dm}^3$$
$$= 0.0836\,\text{mol/dm}^3$$

Step 4 Mass concentration (g/dm^3) = M_r × concentration

$$= 98\,\text{g/mol} \times 0.0836\,\text{mol/dm}^3$$
$$= 8.19\,\text{g/dm}^3$$

examiner tip

Use the same number of significant figures in your answers as are in the question.

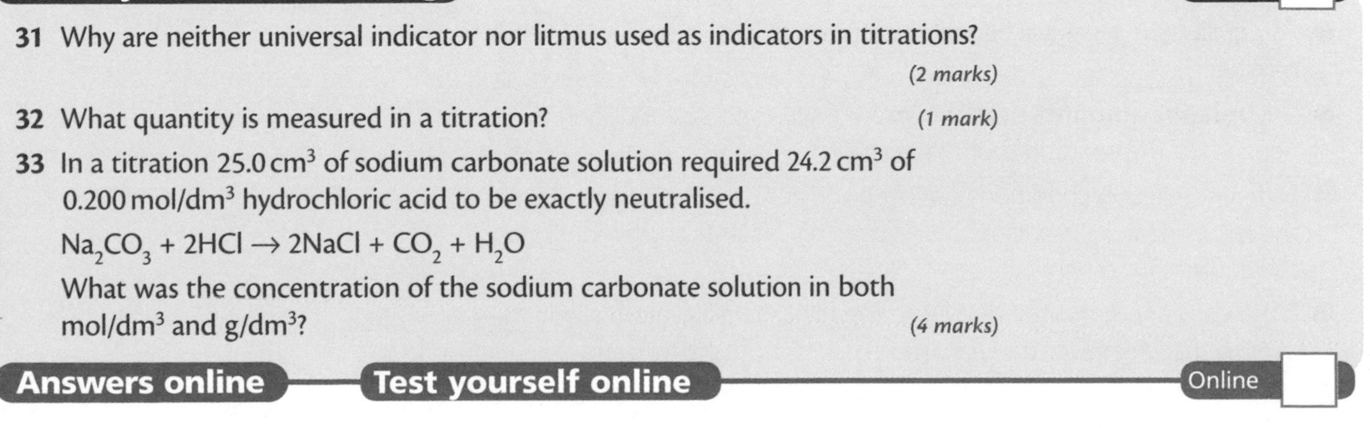

Check your understanding — Tested

31 Why are neither universal indicator nor litmus used as indicators in titrations? *(2 marks)*

32 What quantity is measured in a titration? *(1 mark)*

33 In a titration 25.0 cm^3 of sodium carbonate solution required 24.2 cm^3 of 0.200 mol/dm^3 hydrochloric acid to be exactly neutralised.

$Na_2CO_3 + 2HCl \rightarrow 2NaCl + CO_2 + H_2O$

What was the concentration of the sodium carbonate solution in both mol/dm^3 and g/dm^3? *(4 marks)*

Answers online **Test yourself online** — Online

Making ammonia (1)

The Haber process

Revised

- Ammonia is manufactured using the Haber process:

  nitrogen + hydrogen ⇌ ammonia

  $$N_2 + 3H_2 \rightleftharpoons 2NH_3$$

- The reaction is **reversible** so ammonia breaks down again into nitrogen and hydrogen.
- The **raw materials** for the Haber process are nitrogen (N_2) and hydrogen (H_2).
- **Nitrogen** is obtained from the **air** (78% N_2) and hydrogen can be obtained from **natural gas** or other sources.

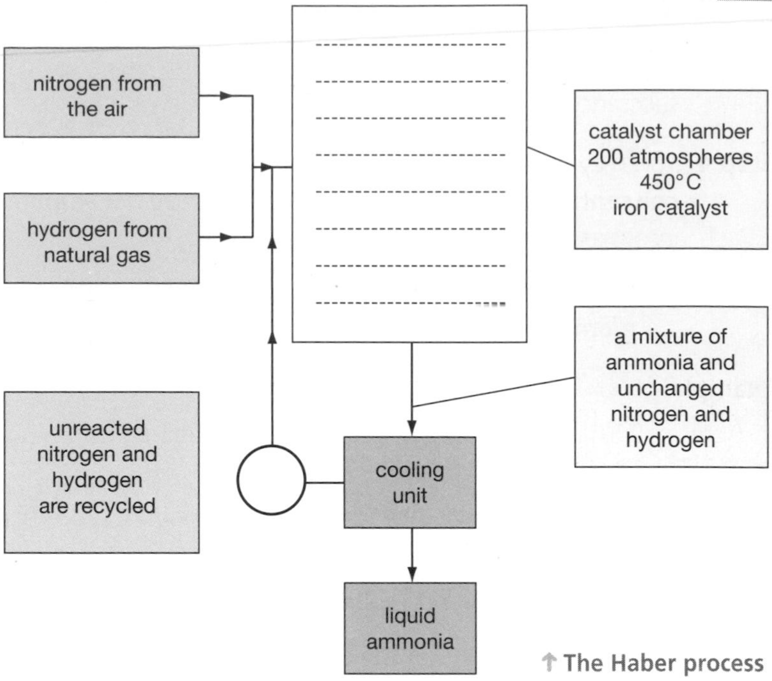

nitrogen from the air

hydrogen from natural gas

catalyst chamber
200 atmospheres
450°C
iron catalyst

a mixture of ammonia and unchanged nitrogen and hydrogen

unreacted nitrogen and hydrogen are recycled

cooling unit

liquid ammonia

↑ The Haber process

There are four, easy-to-learn steps in the Haber process:

1. The **purified gases** are passed over an **iron catalyst** at a **high temperature** (about 450°C) and a **high pressure** (about 200 atmospheres).
2. **Some** of the hydrogen and nitrogen **reacts** to form ammonia.
3. On **cooling**, the **ammonia liquefies** and is removed.
4. The remaining hydrogen and nitrogen are **recycled** (into the catalyst chamber).

Factors affecting equilibrium

Revised

- When a reversible reaction occurs in a **closed** system, nothing can escape and **equilibrium** can be reached.
- This is when the reactions occur at exactly the **same rate in each direction**.
- As fast as the forward reaction is making products, these products are reacting to make the reactants.
- At equilibrium there will be some reactants, as well as products, in the mixture.
- The **relative amounts** of all the reacting substances at equilibrium depend on the conditions of the reaction.
- Different reaction conditions can push the position of equilibrium to the left or right — changing the amount of reactants and products in the equilibrium mixture.
- If the equilibrium position is over to the right there are more products; over to the left means there are more reactants in the mixture.
- **Temperature** conditions affect all reversible reactions; if the reaction involves gases**, pressure** is also important.

examiner tip

Remember: equilibrium can be achieved in a closed system when forward and reverse reactions occur at the same rate.

Effect of temperature on equilibrium

Revised

The table gives a summary of the effects of temperature on the position of equilibrium (the relative amounts of reactants and products).

Change made at equilibrium	If the reaction is...	Effect of change
Increased temperature	Exothermic	Yield decreased
	Endothermic	Yield increased
Decreased temperature	Exothermic	Yield increased
	Endothermic	Yield decreased

Remember that thermal decomposition reactions are endothermic (heat must be added) — this will help you see why increasing the temperature of endothermic reactions increases the **yield**.

examiner tip

Keep it simple!

- Just remember: increased temperature, exothermic direction, yield falls. Look how the table shows the pattern.
- Ignore the rate effects of temperature and pressure when answering **general** equilibrium questions.

Effect of pressure on equilibrium

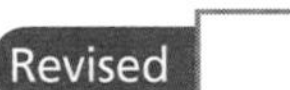

Change made at equilibrium	Effect of change
Increased pressure	Increasing pressure favours the reaction direction that leads to a **reduction in the number of gas molecules**. The amount of these products formed increases. (See the Haber process on page 86.)

Check your understanding

Tested

34 What are the reaction conditions for the Haber process? *(3 marks)*

35 How is the ammonia removed? *(2 marks)*

36 What is done with unreacted nitrogen and hydrogen? *(1 mark)*

37 Under what conditions can a reversible reaction reach equilibrium? *(1 mark)*

38 The decomposition of calcium carbonate is endothermic:

$CaCO_3 \rightleftharpoons CaO + CO_2$

What is the effect of increasing the reaction temperature? *(1 mark)*

39 The oxidation of sulfur dioxide to sulfur trioxide is a gaseous reaction:

$2SO_2(g) + O_2(g) \rightleftharpoons 2SO_3(g)$

What is the effect of increasing the reaction pressure? *(1 mark)*

Answers online **Test yourself online** Online

Making ammonia (2)

Optimum conditions for the Haber process

Revised

Although reversible reactions like the Haber process may not go to completion, they can still be used efficiently in continuous industrial processes.

The reaction conditions have to be chosen carefully to produce a reasonable yield of ammonia quickly.

- The factors that affect equilibrium, together with reaction rates, are important when determining the optimum conditions in industrial processes, including the Haber process.

The forward reaction to produce ammonia is exothermic, **so increasing the temperature** in the Haber process **decreases the yield** of ammonia (see the top table on page 85). It would seem sensible to use a **low temperature**, but that leads to a **slow** reaction.

In gaseous reactions, such as the Haber process, an increase in pressure will favour the reaction that produces the lowest number of molecules, as shown by the equation for that reaction.

Look at the equation again:

$$N_2(g) + 3H_2(g) \rightleftharpoons 2NH_3(g)$$

There are four molecules on the left-hand side and two on the right.

Increasing pressure favours the right-hand side and thus **increases the yield** of ammonia (see page 85). But high-pressure reactors are **expensive** to make, increase the **risk of explosion** and require a **high energy input.**

When answering general questions about equilibria it is best to ignore rate effects. But in the Haber process rate is all important. The ammonia has to be produced as quickly as possible.

- **Reaction rates**, as well as the effect on equilibrium position, are important when determining the optimum conditions in industrial processes.

In the Haber process, **optimum conditions** to produce a **reasonable yield** of ammonia **quickly** are achieved by adjusting the reaction conditions as follows:

1. A catalyst is used to speed up the reaction.
2. High pressure drives the equilibrium forwards, but not so high as to need expensive equipment.
3. A low temperature would increase the yield but make the reaction too slow, so the chosen temperature and pressure combination is a **compromise**.

examiner tip

A catalyst speeds up a reaction — it has **no** effect on yield.

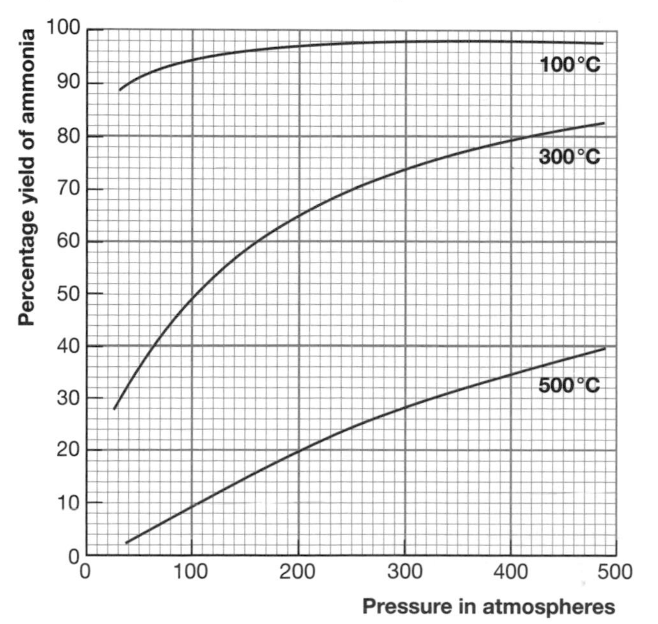

The manufacture of ammonia involves a compromise between high yield and high rate of reaction

Sustainable chemistry in industry

Revised

It is important for **sustainable development** as well as **economic** reasons to:

- minimise **energy requirements**
- minimise **energy wasted** in industrial processes

The Haber process consumes an enormous amount of energy due to the high temperature and pressure conditions.

In other industrial processes, low temperature and pressure conditions (called **non-vigorous**) can be used. This means:

- less **energy is used**
- less **energy is released** into the environment

Running processes sustainably includes the following:

- using catalysts
- recycling unused reactants back into the process
- converting waste products into useful materials

You could be asked to evaluate the conditions used in industrial processes in terms of energy requirements. Make sure you know the effects of temperature and pressure on reactions.

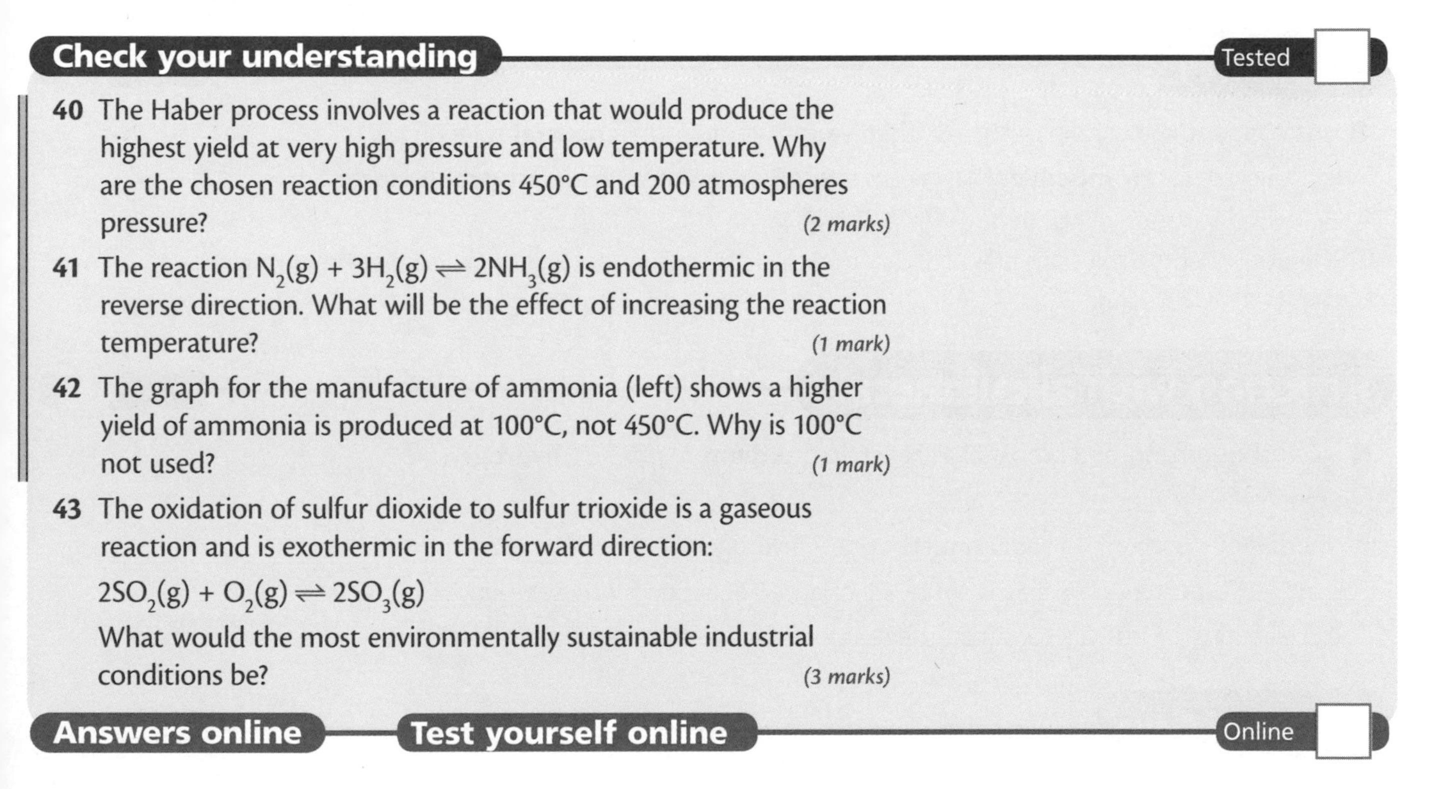

Check your understanding

Tested

40 The Haber process involves a reaction that would produce the highest yield at very high pressure and low temperature. Why are the chosen reaction conditions 450°C and 200 atmospheres pressure? *(2 marks)*

41 The reaction $N_2(g) + 3H_2(g) \rightleftharpoons 2NH_3(g)$ is endothermic in the reverse direction. What will be the effect of increasing the reaction temperature? *(1 mark)*

42 The graph for the manufacture of ammonia (left) shows a higher yield of ammonia is produced at 100°C, not 450°C. Why is 100°C not used? *(1 mark)*

43 The oxidation of sulfur dioxide to sulfur trioxide is a gaseous reaction and is exothermic in the forward direction:

$2SO_2(g) + O_2(g) \rightleftharpoons 2SO_3(g)$

What would the most environmentally sustainable industrial conditions be? *(3 marks)*

Answers online **Test yourself online** Online

Alcohols

A homologous series

Revised

- Alcohols contain the **functional group** –OH.

A 'functional group' is a part of a molecule that takes part in most of its reactions.

The –OH group in alcohols makes them much more **reactive** than the **alkanes**.

- Methanol, ethanol and propanol are the first three members of the **homologous** series of alcohols.

You should be able to recognise any alcohol from the name or formula.

You only need to know the individual names of **methanol, ethanol** and **propanol**.

examiner tip

Compounds in a homologous series have the same general formula. For example, alkanes are C_nH_{2n+2}; alcohols are $C_nH_{2n+1}OH$.

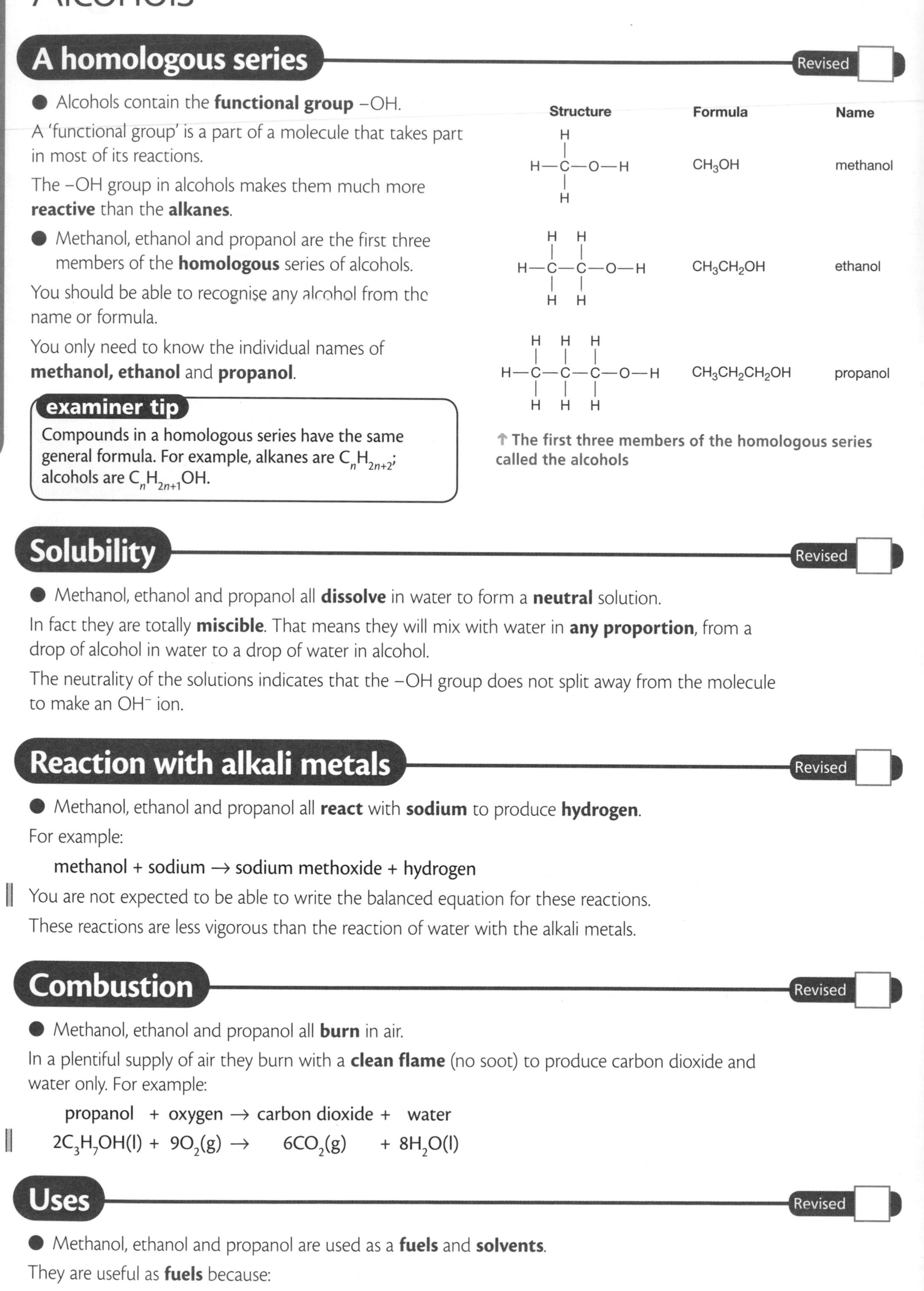

Structure	Formula	Name
H–C(H)(H)–O–H	CH_3OH	methanol
H–C(H)(H)–C(H)(H)–O–H	CH_3CH_2OH	ethanol
H–C(H)(H)–C(H)(H)–C(H)(H)–O–H	$CH_3CH_2CH_2OH$	propanol

↑ The first three members of the homologous series called the alcohols

Solubility

Revised

- Methanol, ethanol and propanol all **dissolve** in water to form a **neutral** solution.

In fact they are totally **miscible**. That means they will mix with water in **any proportion**, from a drop of alcohol in water to a drop of water in alcohol.

The neutrality of the solutions indicates that the –OH group does not split away from the molecule to make an OH^- ion.

Reaction with alkali metals

Revised

- Methanol, ethanol and propanol all **react** with **sodium** to produce **hydrogen**.

For example:

methanol + sodium → sodium methoxide + hydrogen

You are not expected to be able to write the balanced equation for these reactions.

These reactions are less vigorous than the reaction of water with the alkali metals.

Combustion

Revised

- Methanol, ethanol and propanol all **burn** in air.

In a plentiful supply of air they burn with a **clean flame** (no soot) to produce carbon dioxide and water only. For example:

propanol + oxygen → carbon dioxide + water

$$2C_3H_7OH(l) + 9O_2(g) \rightarrow 6CO_2(g) + 8H_2O(l)$$

Uses

Revised

- Methanol, ethanol and propanol are used as a **fuels** and **solvents**.

They are useful as **fuels** because:

- they are liquids that are easily transported and stored
- they burn with clean flames (see the equation above)
- they release a lot of heat energy when they burn

When ethanol is used as a fuel it often has methanol and other substances in it and is called 'meths'. This is used in spirit burners for camping and in food warmers.

'Bioethanol' is made from plant material and is now used as a fuel on its own or as an additive to petrol.

They are useful as **solvents** because they dissolve:

- oil and grease, so are useful as **cleansers**
- inks and pigments, so are useful in **pens** and **paints**
- **perfumes**
- **flavourings**

Alcoholic drinks

Revised

- Ethanol is the alcohol in alcoholic drinks.

It is ethanol's ability to **dissolve** flavouring compounds that gives alcoholic drinks their unique flavours. Water is unable to do this.

Ethanol is **toxic,** but when taken in moderation the body can deal with it.

Ethanol quickly affects the central nervous system, **slowing** your reactions and reducing your ability to make decisions.

examiner tip

When people use the word 'alcohol' they usually mean 'ethanol'.

You must use the correct names of alcohols in chemistry.

Oxidising alcohols

Revised

- Compete oxidation occurs when alcohols burn.
- **Partial oxidation** can be done by using chemical reagents or controlling the supply of oxygen.
- Ethanol can be oxidised to ethanoic acid, either by chemical oxidising agents or by microbial action.

A typical oxidation to ethanoic acid is:

$$\text{ethanol + oxygen} \xrightarrow{\text{acid catalyst}} \text{ethanoic acid + water}$$

You are not expected to be able to write the balanced equation for these reactions.

- Ethanoic acid is the main acid in vinegar.

Alcoholic drinks (wine, beer, cider etc.) turn sour if they are exposed to air for some time, for example overnight.

This sourness is due to ethanoic acid that has been produced by the slow reaction of oxygen in the air with the ethanol in the drink.

Check your understanding

Tested

44 What is a 'functional group'? *(1 mark)*

45 Why are alcohols more reactive than alkanes? *(2 marks)*

46 Give the word equation for the reaction of propanol with sodium. *(2 marks)*

47 For the combustion of ethanol, write:

a) the word equation *(1 mark)*

b) the balanced chemical equation *(2 marks)*

48 Which alcohol is the one in alcoholic drinks? *(1 mark)*

49 Which acid is the main one in vinegar? *(1 mark)*

Answers online **Test yourself online** Online

Carboxylic acids

A homologous series

Revised

- Ethanoic acid is a member of the carboxylic acids, which have the **functional group** –COOH.

The –COOH group in these acids makes them much more **reactive** than the **alkanes**.

- Methanoic acid, ethanoic acid and propanoic acid are the first three members of the **homologous** series of carboxylic acids.

You should be able to recognise any carboxylic acid from the name or formula.

You only need to know the individual names of **methanoic acid**, **ethanoic acid** and **propanoic acid**.

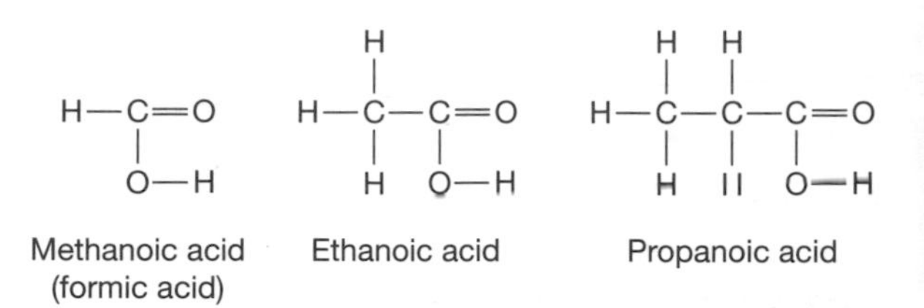

↑ The first three members of the homologous series called the carboxylic acids

Dissolving in water

Revised

- Carboxylic acids dissolve in water to produce acidic solutions.
- Carboxylic acids **do not ionise completely** when dissolved in water and so are **weak** acids.

If 1 mole of ethanoic acid is dissolved in water to make 1 dm^3 of solution, only a few of the acid molecules ionise, as in this equation.

- Aqueous solutions of **weak acids** have a **higher pH** value than aqueous solutions of **strong acids** with the same concentration.

This is because there are only a **few** H^+ ions in the weak acid solution.

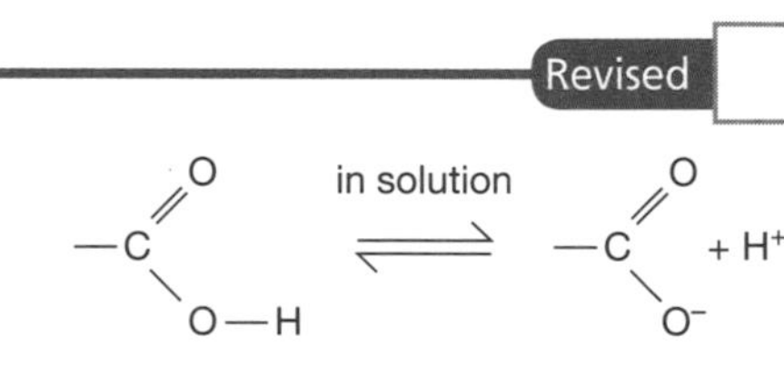

↑ The carboxylic acid functional group –COOH. It forms H^+ ions in solution

Reactions with carbonates

Revised

- Carboxylic acids react with **carbonates** to produce **carbon dioxide**. A salt and water are also produced.

The reactions are **slow**: the acids are **weak**.

The few hydrogen ions that are in solution are used up by the reaction but the functional group equilibrium (see above) shifts to produce more hydrogen ions. This keeps happening until all the acid molecules have been ionised. For example:

ethanoic acid + potassium carbonate → potassium ethanoate + carbon dioxide + water

You are not expected to be able to write the balanced equation for these reactions.

Names of carboxylic acid salts

Acid name	Name of compound produced in reactions
Methanoic	Methanoate
Ethanoic	Ethanoate
Propanoic	Propanoate
Butanoic	Butanoate

examiner tip

Remember: weak acids react more slowly than strong acids.

Reactions with alcohols

Revised

- Carboxylic acids react with **alcohols** in the presence of an acid catalyst to produce **esters.**

The catalyst is often concentrated sulfuric acid (c.H_2SO_4).

For example:

ethanoic acid + propanol $\xrightarrow{c.H_2SO_4}$ propyl ethanoate + water

Esters

A homologous series

Revised

- Esters are a series of sweet-smelling organic compounds, many of which are found in fruits.
- They are **volatile** compounds with **distinctive smells** and are used as **flavourings** and **perfumes**.
- Esters have the **functional group** –COO–.

↑ The ester functional group

Making esters

Revised

- **Ethyl ethanoate** is the ester produced from **ethanol** and **ethanoic acid**.

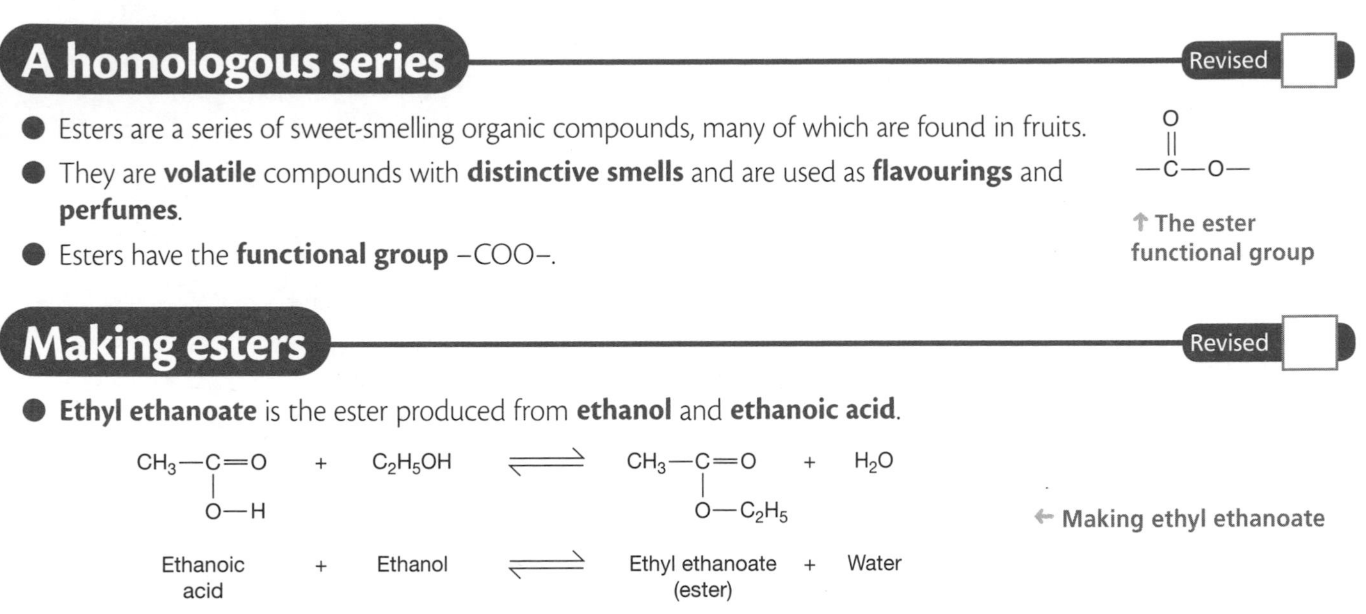

← Making ethyl ethanoate

The name of the ester is a combination of the acid and alcohol from which it can be made.

The naming of three esters

Alcohol	Acid	Ester
Methanol	Ethanoic	Methyl ethanoate
Ethanol	Propanoic	Ethyl propanoate
Propanol	Methanoic	Propyl methanoate

examiner tip

Learn the reaction this way round:

alcohol + acid → ester + water

methanol + **propano**ic acid → **methyl propano**ate + water

Structural formulae

Revised

You need to be able to recognise a compound as an ester from its **name** or its **structural formula**. These structural formulae are distinctive in that they have an easy-to-see branch.

Ester	Structural formula
Methyl ethanoate	$CH_3—C(=O)—O—CH_3$
Ethyl propanoate	$CH_3—CH_2—C(=O)—O—C_2H_5$
Propyl methanoate	$H—C(=O)—O—C_3H_7$

→ The structural formulae of the esters in the 'name' table

Check your understanding

Tested

50 What functional group is in carboxylic acids? *(1 mark)*

51 Why are carboxylic acids acidic in aqueous solution? *(2 marks)*

52 Give the word equation for the reaction of methanoic acid with sodium carbonate. *(2 marks)*

53 Write the word equation for the reaction of propanoic acid with methanol. *(2 marks)*

54 Why are carboxylic acids weak acids? *(2 marks)*

55 How does the pH of a propanoic acid solution compare with that of a nitric acid solution of the same concentration? *(1 mark)*

56 What ester is produced when methanoic acid reacts with ethanol? *(1 mark)*

Answers online **Test yourself online** Online

Index

F

G

H

I

L

M

N

O

Free Time and Leisure

Revised

Education and Work

Revised

The Wider World

Revised

Exam Advice

Listening and Reading

All instructions and questions in both reading and listening tests are now in English. Each test is worth 20% of the total marks, making 40% for both skills combined. At foundation tier, examiners have to set papers using the word lists printed in the specification. At higher tier, there are additional words to learn and some unfamiliar words may be used. You should be able to work out the meaning of these new words using communication strategies (see page 5). Many of the questions will be multiple choice, but other tasks – such as true / false / not mentioned (in reading only), matching exercises, gap fill and answers in English – will also be used. Here are some tips for tackling these comprehension questions:

- Keep revising vocabulary as often as you can. Generally, the more words you know, the easier the tests will be.
- In listening tests, you are given 5 minutes before the CD starts to read through the questions. Use this time sensibly and make notes on the paper if necessary.
- Read the questions carefully. You would be surprised how often people misread questions. For example, in answer to the question 'Apart from football, what other sports does she like?' it is not uncommon to find candidates writing 'football' as their answer.
- Do not panic if you do not understand everything you hear and every single word you read. Working out the gist is the key skill.
- Never leave a blank space. If you are genuinely stuck, make a sensible guess.
- Make sure you form letters clearly and write legibly. If you do decide to cross out an answer, write the new answer as near as possible to the original and make it absolutely clear what your final answer is. Always use a black pen. Your papers will be scanned and marked online.

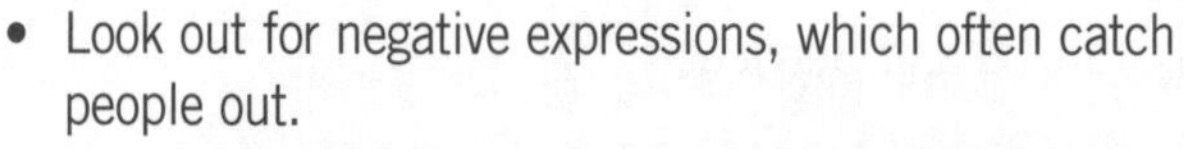

- Look out for negative expressions, which often catch people out.
- Do not assume the first word you see or hear is the one you need for your answer. **No voy al colegio en autobús, voy a pie** means the speaker goes to school on foot and not by bus.
- Recognition of tenses is important. Look at or listen to verb endings with great care.
- High frequency words such as **a menudo**, **siempre**, **nunca**, **salvo** and **otra vez** can change the meaning of a sentence. **Va a la piscina todos los días salvo el domingo** means that s/he does not go to the swimming pool every day, despite the phrase **todos los días**, since **salvo** means 'except'.
- When revising vocabulary, try to learn words in categories so that you can see the connections between words. If there is a question about a person's favourite leisure activity, the answer could be 'reading' but the word **la lectura** (reading) may not be used at all. Terms such as **una novela** (a novel) or **un periódico** (a newspaper) might be used instead.

✓ Maximise Your Marks

In listening and reading exams, always read the question carefully so that you know exactly what to look or listen out for. If the question asks what a person did in the past, look out for time indicators that refer to the past (**hace una semana**, etc.), as well as verbs in the past. At higher tier, there will often be a combination of tenses within the same text or passage, so make sure you identify the exact part that you need.

Communication Strategies

You may have to use communication strategies to work out the meaning of some words used in higher tier papers. You may know the verb **cantar** (to sing), in which case it is reasonable for you to work out, by using your knowledge of the verb, along with other clues and context from the rest of the text, that **cantante** means 'singer' and that **una canción** means 'a song'.

✓ Maximise Your Marks

Look out for time markers to help you decide what tense is being used:

todos los días / normalmente	→	present tense
mañana / la semana próxima	→	future tense
ayer	→	preterite
en el pasado / antes	→	imperfect

Speaking

The speaking component is worth 30% of the total marks and is tested through controlled assessments. You will have to submit marks for two pieces of speaking that you have done during your GCSE course. These will be marked by your teacher. Each assessment lasts about 4–6 minutes. You will be given a task by your teacher and you can then plan what you want to say. You are allowed to have a sheet with up to 30 or 40 words* on it to guide you as you do the task.

- Be aware that at the end of the test your teacher will ask an additional question that you will not have prepared. The answer does not need to be a long one but should contain a verb.
- Make sure you always give plenty of opinions during your talk.
- Include a variety of time frames, past, present and future, showing a good variety of tenses.
- Some of the marks are awarded for pronunciation, so try really hard to sound as Spanish as you can! Take care with certain sounds such as '**j**' – remember, this is not a hard 'j' as in English, but more of a 'h'. So **juego** (I play) is pronounced '*hwaygo*'.
- Avoid answers that are very brief. Always try to develop what you say.
- Make your sentences longer by including connectives such as **porque**, **que**, **donde** and **sin embargo**.

*The exact number of words depends on your exam board. Check this with your teacher.

Boost Your Memory

To help you speak more fluently, why not record yourself saying some general key phrases, such as opinion phrases **creo que** (I think that), **en mi opinión** (in my opinion), etc., or time expressions such as **la semana pasada** (last week), **normalmente** (usually), etc., and then listen to them as often as possible? That way, they will become very familiar to you and by exam time you will be able to use them without needing to try too hard to remember them.

Writing

The writing component is also worth 30% of the total marks and is tested through controlled assessments. You will have to submit two pieces of writing that you have done during your GCSE course. They will be marked by the exam board. Each assessment should be between 100 and 350 words*. You will be given a task by your teacher and you can then plan what you want to write. You are allowed to have a dictionary plus a sheet with up to 30 or 40 words* on it to guide you as you do the task.

- Use a variety of tenses and give plenty of personal opinions and reasons.
- Make sure you use your dictionary to check spellings and genders. Avoid the temptation to look up new words – you can end up making errors, such as writing '**un fósforo de fútbol**'. This does *not* mean a football match, since **un fósforo** is a match you strike to get fire. The Spanish for a football match is **un partido de fútbol**.
- Longer, more complex sentences always score more marks.
- Make use of adverbs, adjectives and pronouns to make your work more personal and interesting.

*The exact number of words depends on your exam board. Check this with your teacher.

✓ Maximise Your Marks

Do not forget to check your work for accuracy, as this is a common way to lose marks in an exam, particularly in the writing section. Check things such as verb endings for the correct person and tense, adjectival agreement, word order, use of accents, etc. This is essential to attain the higher grades.

Basic Phrases and Expressions

Greetings

Hola	Hi / Hello
Buenos días	Good day / Good morning
Buenas tardes	Good afternoon
Buenas noches	Good evening / Good night
¿Qué tal?	How are things?
¿Cómo estás?	How are you? (informal)
¿Cómo está usted?	How are you? (formal)
(Muy) bien, gracias	(Very) well, thanks
Mal / No muy bien	Not very well

Days, Months and Dates

lunes	Monday
martes	Tuesday
miércoles	Wednesday
jueves	Thursday
viernes	Friday
sábado	Saturday
domingo	Sunday
enero	January
febrero	February
marzo	March
abril	April
mayo	May
junio	June
julio	July
agosto	August
septiembre	September
octubre	October
noviembre	November
diciembre	December

This is how to write dates in Spanish:

22 de abril de 2011	the 22nd of April 2011
Llego el 4 de mayo.	I'm arriving on the 4th of May.

¿A qué fecha estamos? What's the date today?

✓ Maximise Your Marks

Except when the first word in a sentence, Spanish days and months do not start with a capital letter as in English. Do not forget to check this in your writing.

Numbers

0	**cero**
1	**uno**
2	**dos**
3	**tres**
4	**cuatro**
5	**cinco**
6	**seis**
7	**siete**
8	**ocho**
9	**nueve**
10	**diez**
11	**once**
12	**doce**
13	**trece**
14	**catorce**
15	**quince**
16	**dieciséis**
17	**diecisiete**
18	**dieciocho**
19	**diecinueve**
20	**veinte**
21	**veintiuno**
22	**veintidós**
30	**treinta**
31	**treinta y uno**
40	**cuarenta**
41	**cuarenta y uno**
50	**cincuenta**
60	**sesenta**
70	**setenta**
80	**ochenta**
90	**noventa**
100	**cien** (exactly 100)
101	**ciento uno**
102	**ciento dos**
200	**doscientos**
500	**quinientos**
700	**setecientos**
900	**novecientos**
1000	**mil**
1000 000	**un millón**

A year is written like this:

1995 **mil novecientos noventa y cinco**

2001 **dos mil uno**

The Alphabet

A	as in 'hat'
B	bay
C	thay
D	day
E	eh
F	efay
G	hay
H	achay
I	ee
J	hota
K	kah
L	elay
M	emay
N	enay
Ñ	enyay
O	oh
P	pay
Q	coo
R	eray
S	essay
T	tay
U	oo
V	oobay
W	oobay doblay
X	ekees
Y	eegreeayga
Z	thayta

Time

¿Qué hora es?	What time is it?
Es la una.	It's one o'clock.
Son las dos.	It's two o'clock.
Son las dos y cuarto.	It's quarter past two.
Son las dos y media.	It's half past two.
Son las dos menos cuarto.	It's quarter to two.
Son las dos y diez.	It's ten past two.
Son las dos menos diez.	It's ten to two.
A las tres	At three o'clock
A la una y veinticinco	At twenty-five past one
Es mediodía.	It's midday.
Es medianoche.	It's midnight.

The Weather

¿Qué tiempo hace?	What's the weather like?
Hace buen tiempo.	It's fine.
Hace mal tiempo.	It's bad weather.
Hace calor.	It's hot.
Hace viento.	It's windy.
Hace sol.	It's sunny.
Hace frío.	It's cold.
Está lloviendo.	It's raining.
Está nevando.	It's snowing.
Está nublado.	It's cloudy.
Está despejado.	There are clear skies.
Hay niebla.	It's foggy.
Hay tormenta.	It's stormy.

? Test Yourself

What do these mean in English?

1. **Son las cuatro y veinte.**
2. **En enero hace frío.**

How do you say these in Spanish?

3. It is half past ten.
4. It is raining and it's foggy.

★ Stretch Yourself

Say or write these in Spanish:

1. I'm arriving on the 21st of January.
2. the 14th of October 2011.

Personal Information

Getting to Know Each Other

Here are some useful questions and answers you could use when talking to others:

¿Cómo te llamas? What is your name?	**Me llamo...** My name is...
¿Cuántos años tienes? How old are you?	**Tengo dieciséis años. Nací en mil novecientos noventa y cinco.** I am 16. I was born in 1995.
¿Cuándo es tu cumpleaños? When is your birthday?	**Es el trece de octubre.** It is the 13th of October.
¿Dónde vives? Where do you live?	**Vivo en Manchester desde hace cinco años. Antes, vivía en Birmingham.** I have been living in Manchester for five years. Before that, I used to live in Birmingham.
¿Cómo eres? What are you like?	**Soy bastante alto con el pelo castaño, corto y liso y los ojos verdes.** I am quite tall with short, straight, dark brown hair and green eyes.
¿Y cómo eres de carácter? And what sort of person are you?	**En mi opinión soy tímido a veces.** In my opinion I am sometimes shy.

Two Useful Verbs

Ser (to be) and **tener** (to have) are very useful verbs to use when describing yourself and others.

Ser is used for describing a permanent quality or state of something. Here is the present tense in full:

Soy I am
Eres You are (**tú**)
Es He / She / It is; You are (**Vd.**)
Somos We are
Sois You are (**vosotros**)
Son They are; You are (**Vds.**)

Soy inglesa. I am English. (feminine)
Sois ingleses. You are English. (masculine plural)

Tener, as well as meaning 'to have', is also used when saying your age. So in Spanish you say 'I *have* 16 years' rather than 'I *am* 16 years old'. Here is the present tense in full:

Tengo I have
Tienes You have **(tú)**
Tiene He / She / It has; You have (**Vd.**)
Tenemos We have
Tenéis You have (**vosotros**)
Tienen They have; You have (**Vds.**)

Tengo quince años. I am 15 (years old).
Tenemos un perro. We have a dog.

Personal Characteristics

¿Cómo eres? What are you like?

Soy..., Es...	I am..., He / She is...
agradable	pleasant
amable	friendly
antipático / a	unfriendly
comprensivo / a	understanding
deportivo / a	sporty
desagradable	unpleasant
estricto / a	strict
egoísta	selfish
extrovertido / a	outgoing
gracioso / a	funny
hablador / ora	talkative
honrado / a	honourable / honest
inteligente	intelligent
perezoso / a	lazy
raro / a	strange
reservado / a	quiet / reserved
serio / a	serious
simpático / a	kind / nice
sincero / a	sincere
tímido / a	shy
travieso / a	naughty

Mi amigo es muy simpático.
My friend is very nice.

Using Adjectives

When you describe something in Spanish, remember to make sure that your adjective (describing word) agrees with the noun (the object being described). The spelling of the adjective changes depending on whether the noun is masculine, feminine, singular or plural.

Here is the pattern for regular adjectives:

Masc. Singular	**bonito**	**amable**	**trabajador**	**azul**
Masc. Plural	**bonitos**	**amables**	**trabajadores**	**azules**
Fem. Singular	**bonita**	**amable**	**trabajadora**	**azul**
Fem. Plural	**bonitas**	**amables**	**trabajadoras**	**azules**
English	pretty	friendly	hard-working	blue

Most adjectives come after the noun that they are describing, for example:

- **Pienso que es una mujer graciosa.**
  I think that she is a funny woman.
- **Mi profesor de inglés es un hombre tímido.**
  My English teacher is a shy man.

However, some adjectives can come before the noun.

With adjectives that come before the noun, the masculine singular form ending in –**o** is shortened. For example, **bueno** becomes **buen**. The feminine form does not change:

- **Es un buen actor.**
  He is a good actor. (masculine)
- **Es una buena amiga.**
  She is a good friend. (feminine)

The following adjectives follow this pattern:

bueno good
malo bad
primero first
tercero third
alguno some / any
ninguno none

Grande is shortened before both a masculine noun and a feminine noun to **gran**. It also changes meaning depending on its position in the sentence. Before a noun it means 'great' (excellent or important) and after a noun it means 'big' (large):

un gran hombre a great man
una gran mujer a great woman
un hombre grande a big man
una mujer grande a big woman

✓ Maximise Your Marks

When using adjectives, always double-check that your adjective agrees with the noun, where necessary, in both gender and number.

Home Life

Build Your Skills: Building More Complex Sentences

Do not talk only about yourself all of the time! Show that you can use different parts of verbs and not just the first person. You could do this within the same sentence, using a conjunction or linking word such as **pero** (but). For example:

- **Diría que soy bastante tímida, pero mi madre dice que soy habladora.**
  I would say that I am quite shy, but my mother says that I am talkative.
- **Mi hermano mayor es extrovertido pero mis padres son reservados.**
  My older brother is outgoing but my parents are reserved.

? Test Yourself

What do these mean in English?

1. **Pienso que somos amables.**
2. **Mi cumpleaños es el veintidós de abril.**

How do you say these in Spanish?

3. I have been living in Glasgow for two years.
4. In my opinion they are kind.

★ Stretch Yourself

Say or write these in Spanish:

1. I would say that I am hard-working but my mother says that I am lazy.
2. In my opinion my English teacher is funny, but my friend says he is strict.

Family and Friends

Friends and Family Members

la familia	the family
la madre	mother
el padre	father
el hermano (mayor)	brother (older)
la hermana (menor)	sister (younger)
el tío	uncle
la tía	aunt
el primo	cousin (male)
la prima	cousin (female)
el abuelo	grandfather
la abuela	grandmother
el sobrino	nephew
la sobrina	niece
el hijo	son
la hija	daughter
el hijo único	only child (male)
la hija única	only child (female)
el padrastro	step-father
la madrastra	step-mother
el hermanastro	step-brother
la hermanastra	step-sister
la mujer / esposa	wife
el marido	husband
los padres	parents
los abuelos	grandparents
el amigo	friend (male)
la amiga	friend (female)
el novio	boyfriend
la novia	girlfriend
el niño	boy
la niña	girl
mi / mis	my (singular / plural)

Boost Your Memory

A lot of the words for family members simply change their endings from masculine to feminine. Make your revision easier by learning them in pairs: for example, **hermano – hermana; tío – tía**, etc. However, be aware that some do not follow this pattern and have to be learnt separately: **el marido – la mujer**.

Describing Family and Friends

¿Cómo es tu hermano / a?
What is your brother / sister like?

Es...	He / She is...
alto / a	tall
bajo / a	short (size)
de estatura mediana	of medium height
gordo / a	fat
delgado / a	thin / slim
bonito / a	nice-looking / pretty
feo / a	ugly
guapo / a	handsome / good-looking
pelirrojo / a	red-haired

Tiene...	He (She) has...
barba	a beard
bigote	a moustache
pecas	freckles
los ojos marrones	brown eyes
el pelo castaño	chestnut brown hair

Lleva gafas. He / She wears glasses.

Tiene el pelo...	He / She has...hair.
corto	short
largo	long
liso	straight
rizado	curly
castaño	chestnut brown
moreno	dark brown
rojo	red
rubio	blonde

Mi hermana es de estatura mediana con el pelo rubio y corto y los ojos verdes.
My sister is of medium height with short blonde hair and green eyes.

Creo que mi hermano es bastante guapo porque tiene los ojos azules.
I think that my brother is quite good-looking because he has blue eyes.

Build Your Skills: Developing Language

Learn to develop your answers to make them longer and more interesting. This will get you more marks than short answers giving simple descriptions. For example:

En mi familia somos seis: mi madre, mi padrastro, mis dos hermanos mayores, mi hermanastro menor y yo. Mi madre es alta y bastante gorda y lleva gafas. Nació en mil novecientos sesenta y cinco. Mis hermanos mayores son los dos muy altos y diría que mi hermanastro es bastante bajo. Mi padrastro tiene barba y bigote y creo que es amable.

There are six of us in my family: my mother, my step-father, my two older brothers, my younger step-brother and myself. My mother is tall and quite fat and wears glasses. She was born in 1965. My older brothers are both very tall and I would say that my step-brother is quite short. My step-father has a beard and moustache, and I think that he is friendly.

Mi mejor amiga se llama María y es muy baja y delgada con el pelo castaño y bastante largo. Tiene los ojos verdes y también tiene pecas. En mi opinión es muy bonita y simpática.

My best friend is called Maria and she is very short and slim with quite long, chestnut brown hair. She has green eyes and she also has freckles. In my opinion she is very pretty and kind.

Intensifiers

Make your speaking and writing more interesting by adding intensifiers:

muy	very	**demasiado**	too
bastante	quite	**un poco**	a little
poco	little	**casi**	almost
no...nada	not at all		

Mi hermano es demasiado reservado.
My brother is too quiet / reserved.

The Verb 'Llamarse'

Here are three parts of the present tense of **llamarse** (to be called):

Me llamo	I am called
Te llamas	You are called **(tú)**
Se llama	He / She / It is called; You are called **(Vd.)**

Subject Pronouns

Subject pronouns (I, you, etc.) are not used as often in Spanish as they are in English. In Spanish, it is the verb ending that shows who is doing the action. Subject pronouns are therefore normally used to add emphasis to a sentence.

There are four ways to say 'you' in Spanish, so try to get the correct one!

- **Tú** is used for a friend or someone younger than you. If there are several of these people, use **vosotros**.
- **Usted** is for someone you do not know or who is older than you. For several of these people, use **ustedes**.

yo	I
tú	you (informal, singular)
él	he / it
ella	she / it
usted	you (formal, singular – abbreviated to **Vd.**)
nosotros / as	we
vosotros / as	you (informal, plural)
ellos	they (masculine, or mixed masculine and feminine)
ellas	they (feminine)
ustedes	you (formal, plural – abbreviated to **Vds.**)

? Test Yourself

Rearrange the words into the correct order:

1. **tiene y pelo el prima rizado mi castaño** (My cousin has chestnut brown curly hair.)
2. **alto los y azules bastante ojos soy tengo** (I am quite tall and I have blue eyes.)

How do you say these in Spanish?

3. My aunt is of medium height. My grandfather is quite short.
4. My grandmother is very slim. My son has freckles.

★ Stretch Yourself

Say or write these in Spanish:

1. In my opinion my brother is quite good-looking because he is tall with green eyes and dark brown hair.
2. My best friend is 15 years old. She has brown eyes and blonde hair was born in 1996.

The Present Tense

Using the Present Tense

We use the present tense to talk about what happens now, e.g. 'I study Spanish at school', and about what usually happens, e.g. 'I eat my lunch every day'.

Spanish verbs are divided into three main groups:

- those ending in **–ar**, e.g. **hablar** (to speak / talk)
- those ending in **–er**, e.g. **comer** (to eat)
- those ending in **–ir**, e.g. **vivir** (to live)

Remember that, in Spanish, you must make sure that you use the correct part of the verb to show who is doing the action. This is because the subject pronouns (I, you, etc.) do not tend to be used as often in Spanish. The only way that you can tell who is doing the action is by looking at the verb ending.

The Present Tense of '–ar' Verbs	
Hablar	To speak
Hablo	I speak
Hablas	You speak (**tú**)
Habla	He / She / It speaks; You speak (**Vd.**)
Hablamos	We speak
Habláis	You speak (**vosotros**)
Hablan	They speak; You speak (**Vds.**)

Hablo francés y español. ¿Y tú?
I speak French and Spanish. And you?

Mis padres hablan inglés y alemán.
My parents speak English and German.

Hablamos todos los días por teléfono.
We speak on the phone every day.

The Present Tense of '–er' Verbs	
Comer	To eat
Como	I eat
Comes	You eat (**tú**)
Come	He / She / It eats; You eat (**Vd.**)
Comemos	We eat
Coméis	You eat (**vosotros**)
Comen	They eat; You eat (**Vds.**)

¿Siempre coméis en el comedor?
Do you always eat in the dining room?

Come a la una y media todos los días.
He eats at half past one every day.

The Present Tense of '–ir' Verbs	
Vivir	To live
Vivo	I live
Vives	You live (**tú**)
Vive	He / She / It lives; You live (**Vd.**)
Vivimos	We live
Vivís	You live (**vosotros**)
Viven	They live; You live (**Vds.**)

Vivimos en el sur de Inglaterra.
We live in the south of England.

Mis amigos viven en Perú.
My friends live in Peru.

Irregular Verbs in the Present Tense

Some verbs are irregular in the first person, the **yo** (I) form. The rest of the verb follows the patterns on page 12:

dar → **doy** (I give)
ver → **veo** (I see)
conducir → **conduzco** (I drive)
hacer → **hago** (I do / make)
salir → **salgo** (I go out)
conocer → **conozco** (I know – person)
poner → **pongo** (I put)
saber → **sé** (I know – fact)
traer → **traigo** (I bring)

Infinitives

Infinitives translate as 'to...' in English, e.g. **ver** (to see), **comer** (to eat), etc. This is the part that you will find in the dictionary whenever you look up a verb. Infinitives are used after phrases such as **me gusta** (I like) and after modal verbs such as **poder** (to be able), **querer** (to want) and **preferir** (to prefer).

Quiero salir con mis tíos el fin de semana.
I want to go out with my aunt and uncle at the weekend.

Puedo charlar con mis padres sobre cualquier tema.
I can chat to my parents about any topic.

Prefiero compartir mi dormitorio con mi hermana.
I prefer to share my room with my sister.

Me gusta jugar al fútbol con mis amigos.
I like to play football with my friends.

Quiero enamorarme de un hombre simpático y comprensivo.
I want to fall in love with a kind, understanding man.

Negatives

To make a verb negative in Spanish, put **no** before the verb:
No hablo francés. I don't speak French.

Sometimes the negative phrase is in two parts and sometimes just one. When in two parts, the negative comes before and after the verb:

Nunca veo la tele por la noche.
No veo nunca la tele por la noche.
I never watch TV at night.

Tampoco leo el periódico.
No leo el periódico tampoco.
I don't read the newspaper either.

Here are some more negative expressions:

no...nadie	no one / nobody
no...nunca / jamás	never
no...ni...ni	neither...nor
no...ninguno	no / not any / none

✓ Maximise Your Marks

When completing listening exercises, listen out for negatives, as they can completely alter the meaning of a sentence. You could be caught out if you do not spot the negative. Also, by adding a negative statement to your speaking or writing, you can extend your sentences and gain more marks.

? Test Yourself

What do these mean in English?

1. **Mi hermana vive en una casa grande.**
2. **Comemos a la una y media.**

How do you say these in Spanish?

3. I speak Spanish, you speak French.
4. I go out with my friends.

★ Stretch Yourself

Say or write these in Spanish:

1. I want to know the date.
2. I never eat in the dining room.

Relationships

Talking About Marital Status

Está...	He / She is...
casado / a	married
soltero / a	single
separado / a	separated
divorciado / a	divorced
jubilado / a	retired
viudo / a	widowed
muerto / a	dead

Mi hermana está casada.
My sister is married.

Mis tíos están divorciados.
My aunt and uncle are divorced.

Useful Verbs to Describe Relationships

casarse	to get married
charlar	to chat
compartir	to share
conocer a	to know (person)
confiar en	to trust
detestar	to hate
discutir	to argue / discuss
divorciarse	to get divorced
enamorarse de	to fall in love with
enfadarse con	to get annoyed with
jubilarse	to retire
llevarse bien con	to get on well with
llevarse mal con	to get on badly with
llorar	to cry
molestar	to bother / annoy
odiar	to hate
parecerse a	to look like / resemble
pelearse	to fight / argue
preocuparse de	to worry about
querer	to love
reír	to laugh
separarse	to separate
sonreír	to smile

The Verb 'Estar'

Estar (to be) is used to describe temporary situations. It is also used to describe marital status. Here is the present tense in full:

Estoy	I am
Estás	You are (**tú**)
Está	He / She / It is; You are (**Vd.**)
Estamos	We are
Estáis	You are (**vosotros**)
Están	They are; You are (**Vds.**)

Prepositions and Disjunctive Pronouns

Disjunctive pronouns are pronouns that follow prepositions. They are the same as the subject pronouns (I, you, etc.) with the exception of **yo** and **tú**, which change to **mí** and **ti**.

mí	me
ti	you (informal, singular)
él	him / it
ella	her / it
usted	you (formal, singular)
nosotros / as	us
vosotros / as	you (informal, plural)
ellos	them (masculine)
ellas	them (feminine)
ustedes	you (plural, formal)

Mí has an accent but **ti** does not. This is to distinguish the pronoun **mí** (me) from the possessive adjective **mi** (my).

Some common prepositions are:

con	with
para	for / to
a	to / at
sin	without
de	from / of
por	through / by

With **con**, the pronouns **mí** and **ti** are joined to the preposition to form one word, like this:

conmigo	with me
contigo	with you

More Adjectives for Describing People

When describing a person's qualities or personality, remember to use **ser** (not **estar**):

Soy... **Es...**	I am... He / She is...
alegre	cheerful
bueno / a	good
débil	weak
dinámico / a	dynamic
duro / a	hard / harsh
estúpido / a	stupid / silly
falso / a	false / insincere
listo / a	clever / smart
malo / a	bad
pesado / a	tedious
responsable	responsible
ruidoso / a	noisy
sabio / a	wise / sensible
severo / a	strict
trabajador / a	hard-working
vago / a	lazy / unreliable

Build Your Skills: Higher-level Adjectives

You may find these adjectives used in higher-tier reading and listening exams. They are good to include in your speaking and writing too!

animado / a	lively
asqueroso / a	nasty
celoso / a	jealous
curioso / a	strange / odd
bien/mal educado / a	well / badly behaved
egoísta	selfish / self-centered
enojoso / a	annoying
exigente	demanding
holgazán / ana	lazy / idle
sensato / a	sensible
sensible	sensitive
terco / a	obstinate

✓ Maximise Your Marks

If asked to describe your family or friends, try not to repeat the same type of information (hair and eye colour, etc.) for each person. Aim to vary the information and use as many different adjectives, verbs and expressions as you can to make your speaking and writing more varied and interesting.

Putting It All Together – Talking About Relationships

¿Te llevas bien con tu familia?
Do you get on well with your family?

Pues, a veces sí, pero también discuto bastante con mis padres sobre los deberes.
Well, yes sometimes, but I also argue quite a lot with my parents about my homework.

¿Y qué tal con tus hermanos?
And what about with your brothers and sisters?

Bueno, me llevo muy bien con mi hermana porque es muy madura y comprensiva, pero mis hermanos son tontos y me enfado mucho con ellos.
Well, I get on well with my sister because she is very mature and understanding, but my brothers are stupid and I get really annoyed with them.

¿A quién te pareces?
Who do you look like?

Me parezco mucho a mi madre, pero de carácter soy más como mi padre – extrovertido y a veces perezoso.
I look a lot like my mother, but in personality I am more like my father – outgoing and sometimes lazy.

¿Cómo son tus amigos?
What are your friends like?

Mi mejor amiga es justa y divertida y siempre habla conmigo de todo.
My best friend is fair and good fun and she always talks to me about everything.

? Test Yourself

What do these mean in English?

1. **Mi abuela está viuda.**
2. **Mi padre es muy divertido.**

How do you say these in Spanish?

3. My brother is single.
4. My grandparents are separated.

★ Stretch Yourself

Say or write these in Spanish:

1. I get on well with my brother but my sister is annoying at times.
2. I don't get on well with my brother because he is selfish.

Radical-changing Verbs

What Are Radical-changing Verbs?

Radical-changing verbs are sometimes known as 'stem-changing verbs' or '1, 2, 3, 6 verbs'. They are verbs which have changes to the vowel(s) in the stem or root of the verb, as well as the normal changes to the endings. In the present tense, the changes occur in all persons of the verb except the 'we' part and the 'you' familiar plural part – hence the name '1, 2, 3, 6 verbs'.

There are three common groups:

- **e > ie** (where an 'e' in the stem changes to 'ie')
- **o > ue** (where an 'o' in the stem changes to 'ue')
- **e > i** (where an 'e' in the stem changes to 'i')

The 'e' > 'ie' Group

Querer To want
Quiero I want
Quieres You want (**tú**)
Quiere He / She / It wants; You want (**Vd.**)
Queremos We want
Queréis You want (**vosotros**)
Quieren They want; You want (**Vds.**)

The following verbs change in this way:

cerrar	to close
comenzar	to start / begin
despertar(se)	to wake (up)
empezar	to begin
nevar	to snow
pensar	to think
sentarse	to sit down
encender	to light
entender	to understand
perder	to lose
tener	to have
divertirse	to amuse oneself / have fun
preferir	to prefer
sentir	to feel
venir	to come

Note that **tener** and **venir** are irregular in the first person singular – **tengo** (I have) and **vengo** (I come).

¿Prefieres el rojo o el azul?
Do you prefer the red one or the blue one?

Prefiero el rojo.
I prefer the red one.

¿Quieres té o café?
Do you want tea or coffee?

Quiero un café con leche.
I want a white coffee

¿A qué hora vienen esta tarde?
What time are they coming this afternoon?

¿A qué hora empieza la película?
What time does the film start?

Nieva mucho en el invierno.
It snows a lot in winter.

¿Entiendes el español?
Do you understand Spanish?

¿Tienen mucho dinero?
Do they have a lot of money?

¡Pensamos que está muy bien!
We think that it is very good!

The 'o' > 'ue' Group

Volver To return
Vuelvo I return
Vuelves You return (**tú**)
Vuelve He / She / It returns; You return (**Vd.**)
Volvemos We return
Volvéis You return (**vosotros**)
Vuelven They return; You return (**Vd.**)

The following verbs change in this way:

acordarse	to remember
almorzar	to have lunch
contar	to tell / count
costar	to cost
encontrar	to meet / find
mostrar	to show
probar	to have a go / try
recordar	to remember
volar	to fly
devolver	to give back / return
doler	to hurt
llover	to rain
poder	to be able to
dormir	to sleep
morir	to die

Note that **jugar** (to play) follows this pattern, changing the **u** to **ue**.

¡Cuéntame lo que pasó!
Tell me what happened!

¡No encuentro las llaves por ningún sitio!
I can't find the keys anywhere!

Normalmente, almuerzo a las dos y media.
I usually have lunch at half past two.

Los helados no cuestan mucho.
The ice-creams don't cost a lot.

Boost Your Memory

To achieve the higher grades, you should demonstrate that you know how to use different parts of verbs as often as possible. Make sure that you learn the endings as well as the root changes. Try saying the endings aloud quickly as a chant over and over in a rhythm!

The 'e' > 'i' Group

Pedir To ask for
Pido I ask
Pides You ask (**tú**)
Pide He / She / It asks; You ask (**Vd.**)
Pedimos We ask
Pedís You ask (**vosotros**)
Piden They ask; You ask (**Vds.**)

The following verbs change in this way:

decir	to say
despedirse de	to say goodbye to
medir	to measure
reír	to laugh
repetir	to repeat
seguir	to follow
servir	to serve
vestirse	to dress oneself

Note that **decir** and **seguir** are irregular in the first person singular – **digo** (I say) and **sigo** (I follow).

El profesor repite la pregunta.
The teacher repeats the question.

¿Te vistes a la misma hora todos los días?
Do you get dressed at the same time every day?

Yo siempre pido postre.
I always order a dessert.

Miden la distancia.
They measure the distance.

Test Yourself

What do these mean in English?

1. **Me divierto mucho con mis amigos.**
2. **¿Cuántas horas duermes?**

How do you say these in Spanish?

3. What do you think? (**tú**)
4. They follow the instructions.

Stretch Yourself

Say or write these in Spanish:

1. What time do you usually get dressed on a school day? (**tú**)
2. The house costs a lot of money.

Daily Routine

Reflexive Verbs

Reflexive verbs describe an action that you do to yourself. This means that the subject of the verb is the same as the object of the verb, i.e. the person doing the action is also receiving the action. For example:

Me lavo I wash (myself)
Te vistes You get (yourself) dressed

Reflexive verbs are formed in the same way as other verbs except that they need a reflexive pronoun before each part to say 'myself', etc.

Here is the present tense of the verb **lavarse** in full:

Lavarse	To wash (oneself) / To get washed
Me lavo	I wash (myself)
Te lavas	You wash (yourself) (**tú**)
Se lava	He / She / It washes (him- / her- / itself); You wash (yourself) (**Vd.**)
Nos lavamos	We wash (ourselves)
Os laváis	You wash (yourselves) (**vosotros**)
Se lavan	They wash (themselves); You wash (yourselves) (**Vds.**)

Here are some useful reflexive verbs to talk about your daily routine:

acostarse	to go to bed
afeitarse	to shave
bañarse	to have a bath
cepillarse	to brush (hair / teeth)
desnudarse	to get undressed
despertarse	to wake up
ducharse	to have a shower
levantarse	to get up
peinarse	to comb your hair
vestirse	to get dressed

Note that **acostarse**, **despertarse** and **vestirse** are radical-changing verbs. They follow these patterns:

acostarse (**o** > **ue**) → **me acuesto**, etc.
despertarse (**e** > **ie**) → **me despierto**, etc.
vestirse (**e** > **i**) → **me visto**, etc.

With the infinitive, the reflexive pronoun is usually added to the end of the verb, making a single word:

Suelo levantarme a las seis.
I usually get (myself) up at six o'clock.

Mi hermana suele ducharse a las siete.
My sister usually has a shower at seven o'clock.

Time Expressions

Different time expressions make your speaking and writing about daily routines more varied and interesting. Try to add time expressions to your work as often as you can!

al mediodía	at midday
a menudo	often
a veces	sometimes
después del colegio	after school
de vez en cuando	from time to time
dos veces a la semana	twice a week
generalmente	generally
los fines de semana	at weekends
los sábados	on Saturdays
normalmente	normally / usually
nunca	never
pocas veces	rarely
por la mañana	in the morning
por la tarde	in the afternoon
por la noche	at night
raramente	rarely
siempre	always
suelo (+ infinitive)	I usually (+ infinitive)
todos los días	every day
últimamente	lately

Talking About Daily Routines

¿Cómo es tu rutina diaria los días de colegio?
What's your daily routine like on a school day?

Pues, normalmente me levanto temprano, sobre las siete, y me ducho.
Well, I usually get up early, about seven o'clock, and I have a shower.

Desayuno, me arreglo y salgo de casa a las ocho y cuarto.
I have breakfast, get myself ready and I leave the house at quarter past eight.

Suelo ir al colegio en autobús.
I usually go to school by bus.

¿Qué haces por la tarde?
What do you do in the afternoon?

A veces, me quedo un rato en el colegio para practicar deporte.
Sometimes I stay at school a while in order to do sport.

Vuelvo a casa sobre las cuatro y media y generalmente hago mis deberes en seguida.
I return home at half past four and generally I do my homework straight away.

Por la noche, suelo usar el ordenador para chatear con mis amigos o navegar por internet.
In the evening, I usually use the computer to chat with my friends or to surf the Internet.

¿Y los fines de semana, tu rutina es distinta?
And is your routine different at weekends?

Sí, ¡a menudo me quedo en la cama hasta tarde!
Yes, I usually stay in bed until late!

Build Your Skills: Extending Sentences

Using a variety of expressions in your writing and speaking, including time expressions, is a good way of making your sentences more complex and extended. This will help you gain the higher grades.

Suelo levantarme temprano y me ducho. Luego, me visto y voy de compras con mis amigos o a veces voy a casa de mi amiga.
I usually get up early and I have a shower. Then, I get dressed and I go shopping with my friends or sometimes I go to my friend's house.

✓ Maximise Your Marks

When speaking or writing, get into the habit of changing tenses within a paragraph or even within a sentence, to demonstrate that you can use past, present and future. This is a necessity when aiming for an A or A*. For example:

- **Suelo levantarme a las siete menos cuarto los días de colegio pero esta mañana no me desperté hasta las ocho. ¡Qué desastre! Tuve que salir de casa sin desayunar y ¡ahora tengo mucha hambre!**
  I usually get up (present) at quarter to seven on a school day but this morning I didn't wake up (past) until eight o'clock. What a disaster! I had to leave (past) home without having breakfast and now I'm (present) really hungry!

? Test Yourself

What do these mean in English?

1. **me peino te vistes nos despertamos**
2. **se despierta os laváis te acuestas**

How do you say these in Spanish?

3. I usually get dressed at half past seven.
4. My brother gets up at quarter to eight.

★ Stretch Yourself

Say or write these in Spanish:

1. I usually go to bed at eleven o'clock but tonight I'm going to go to bed at ten.
2. Every day I get up at seven o'clock but on Saturdays I stay in bed until nine.

House and Home

Describing Your House

una casa	house
una casa adosada	semi-detached house
un chalé / un chalet	detached house
una granja	farm
un piso	flat
el árbol	tree
el césped	lawn
la cocina	kitchen
el comedor	dining room
el cuarto	room
el cuarto de estar	living room
el cuarto de baño	bathroom
el desván	attic / loft
el dormitorio	bedroom
la escalera	stairs
el estudio	study
el garaje	garage
la habitación	room
el invernadero	greenhouse
el jardín	garden
el lavadero	utility room
la pared	wall
el pasillo	corridor
el patio	patio
la planta baja	ground floor
la primera planta	first floor
la sala de estar	sitting room
el salón	lounge
el sótano	basement
la terraza	patio / terrace
abajo	downstairs
arriba	upstairs

Vivo en una casa adosada con tres habitaciones en la planta baja. En la primera planta hay tres dormitorios y un cuarto de baño.
I live in a semi-detached house with three rooms on the ground floor. On the first floor there are three bedrooms and a bathroom.

Detrás de la casa hay un jardín bastante grande con un césped y muchas flores. También tenemos un garaje y un invernadero.
Behind the house there is quite a big garden with a lawn and lots of flowers. We also have a garage and a greenhouse.

Describing Where Things Are – Prepositions

When describing the location of a person or thing, use the verb **estar** with a prepostion:

Está / Están...	It Is / They are...
al final de	at the end of
al lado de	next to
a la derecha de	to the right of
a la izquierda de	to the left of
debajo de	below / under
delante de	in front of
detrás de	behind
encima de	above / on
enfrente de	opposite
entre	in between
en el suelo	on the floor
en medio de	in the middle of
en	in / on

Remember: **de + el = del**
de + la = de la

Mi casa está al final de la calle.
My house is at the end of the street.

El garaje está a la derecha de la casa.
The garage is to the right of the house.

La cocina está al lado del salón.
The kitchen is next to the lounge.

Home Life

Describing Your Bedroom

el armario	wardrobe
el cajón	drawer
el despertador	alarm clock
el edredón	duvet
el espejo	mirror
el estante	shelf
el estéreo	hi-fi system
los muebles	furniture
el ordenador	computer
el radiador	radiator
el secador	hairdryer
el techo	ceiling
el tocador	dressing table
la alfombra	rug
la almohada	pillow
la cama	bed
la cómoda	chest of drawers
las cortinas	curtains
la lámpara	lamp
la luz	light
la manta	blanket
la moqueta	carpet
la puerta	door
la sábana	sheet

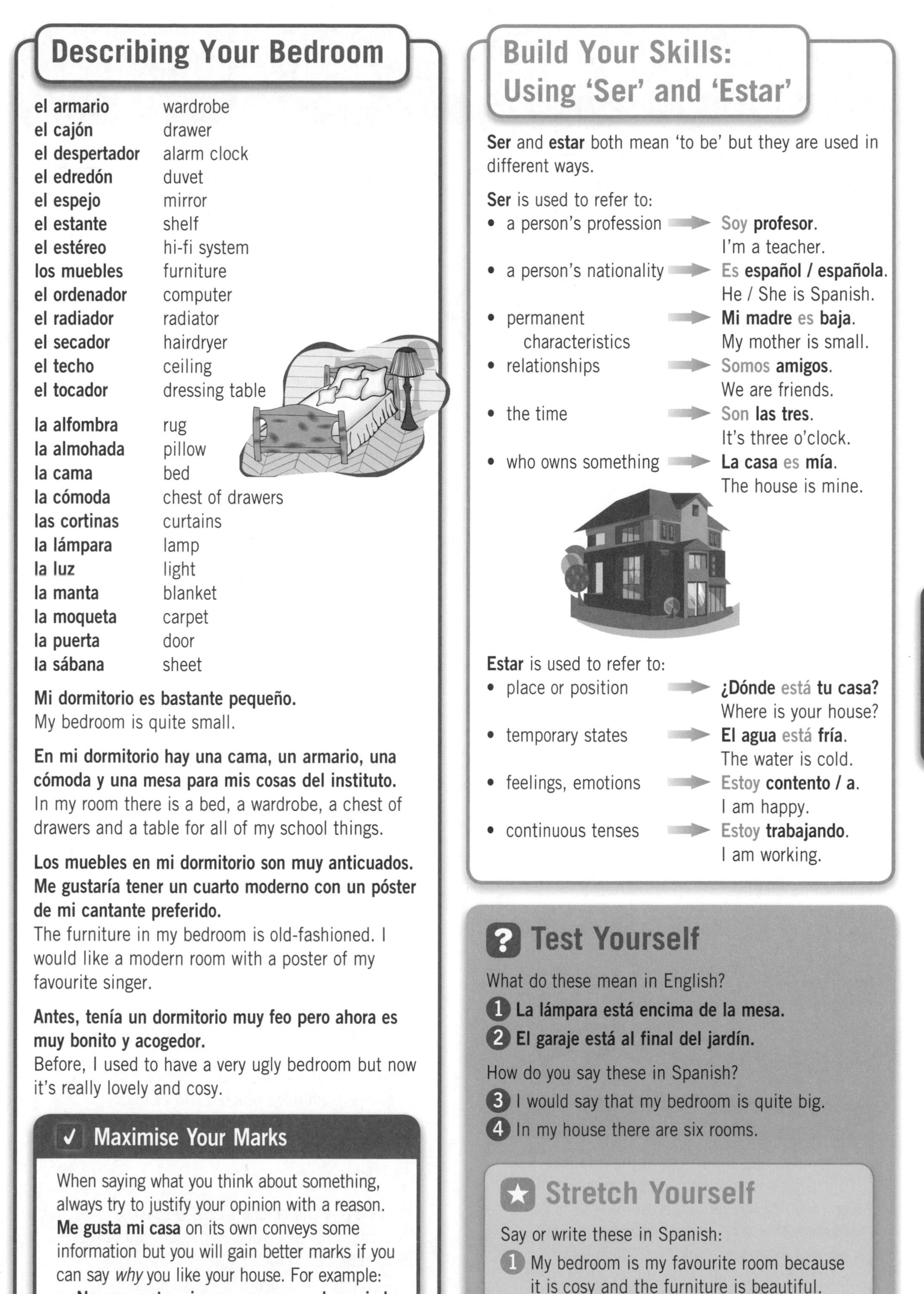

Mi dormitorio es bastante pequeño.
My bedroom is quite small.

En mi dormitorio hay una cama, un armario, una cómoda y una mesa para mis cosas del instituto.
In my room there is a bed, a wardrobe, a chest of drawers and a table for all of my school things.

Los muebles en mi dormitorio son muy anticuados. Me gustaría tener un cuarto moderno con un póster de mi cantante preferido.
The furniture in my bedroom is old-fashioned. I would like a modern room with a poster of my favourite singer.

Antes, tenía un dormitorio muy feo pero ahora es muy bonito y acogedor.
Before, I used to have a very ugly bedroom but now it's really lovely and cosy.

✓ Maximise Your Marks

When saying what you think about something, always try to justify your opinion with a reason. **Me gusta mi casa** on its own conveys some information but you will gain better marks if you can say *why* you like your house. For example:

- **No me gusta mi casa porque es demasiado pequeña y no tenemos jardín.**
  I don't like my house because it is too small and we don't have a garden.

Build Your Skills: Using 'Ser' and 'Estar'

Ser and **estar** both mean 'to be' but they are used in different ways.

Ser is used to refer to:

- a person's profession → **Soy profesor.** I'm a teacher.
- a person's nationality → **Es español / española.** He / She is Spanish.
- permanent characteristics → **Mi madre es baja.** My mother is small.
- relationships → **Somos amigos.** We are friends.
- the time → **Son las tres.** It's three o'clock.
- who owns something → **La casa es mía.** The house is mine.

Estar is used to refer to:

- place or position → **¿Dónde está tu casa?** Where is your house?
- temporary states → **El agua está fría.** The water is cold.
- feelings, emotions → **Estoy contento / a.** I am happy.
- continuous tenses → **Estoy trabajando.** I am working.

? Test Yourself

What do these mean in English?

1. **La lámpara está encima de la mesa.**
2. **El garaje está al final del jardín.**

How do you say these in Spanish?

3. I would say that my bedroom is quite big.
4. In my house there are six rooms.

★ Stretch Yourself

Say or write these in Spanish:

1. My bedroom is my favourite room because it is cosy and the furniture is beautiful.
2. The garden is behind the house.

Helping at Home

What's in Your House?

el abrelatas	tin-opener
el cepillo de dientes	toothbrush
el champú	shampoo
el cojín	cushion
el congelador	freezer
el cubo	bucket
el fregadero	sink
el frigorífico	fridge
el grifo	tap
el horno	oven
el jabón	soap
el lavabo	washbasin
el lavaplatos	dishwasher
el maquillaje	make-up
el microondas	microwave
el plato	plate / dish
el sacacorchos	corkscrew
la aspiradora	vacuum cleaner
la bandeja	tray
la bombilla	light bulb
la butaca	armchair
la calefacción central	central heating
la chimenea	fireplace
la cocina eléctrica / de gas	cooker (electric / gas)
la ducha	shower
la lavadora	washing machine
la nevera	fridge
la papelera	waste-paper bin
la pasta de dientes	toothpaste
la sartén	frying pan
la silla	chair
la taza	cup
las tijeras	scissors
la toalla	towel

Boost Your Memory

When learning a new verb in the infinitive, it is a good idea to immediately practise conjugating the verb, at least into the first person. That way, as well as knowing that **cocinar** means 'to cook', you can also say **cocino** (I cook) and maybe also **cociné** (I cooked). If you get into the habit of doing this, it will improve your speaking and writing skills as well as your comprehension, especially with irregular verbs such as **hacer** → **hago**.

Useful Verbs for Helping at Home

apagar (la luz)	to switch off (light)
calentar	to heat
cerrar	to close
cocinar	to cook
coger	to get
encender (la luz)	to switch on (light)
limpiar	to clean
preparar	to prepare
reparar	to repair
secar	to dry
barrer el suelo	to sweep the floor
cortar el césped	to cut the lawn
fregar los platos	to do the washing-up
hacer de canguro	to babysit
hacer la comida	to do the cooking
hacer las compras	to do the shopping
hacer mi cama	to make my bed
lavar la ropa	to do the washing / wash the clothes
ordenar mi dormitorio	to tidy my room
pasar la aspiradora	to vacuum
pasear al perro	to walk the dog
planchar	to iron
poner la mesa	to lay the table
quitar la mesa	to clear the table
sacar la basura	to put the rubbish out
los quehaceres / las faenas	chores

¿Tienes que ayudar en casa?
Do you have to help with the housework?

Sí, tengo que fregar los platos todos los días.
Yes, I have to do the washing-up every day.

A veces mi hermano saca la basura o pone la mesa pero no hace mucho. ¡No es justo!
Sometimes my brother puts the rubbish out or sets the table, but he doesn't do a lot. It's not fair!

Anoche planché la ropa y barrí el suelo. No me gusta nada planchar porque creo que es muy aburrido.
Last night I ironed the clothes and I swept the floor. I don't like ironing at all because I think that it's really boring.

Build Your Skills: Using 'Tener Que'

To say you *have to* do something, use **tener que** + infinitive. Remember to use different parts of the verb and different tenses as well. For example:

- Tengo que **quitar la mesa todos los días y mi hermano** tiene que **fregar los platos**.
  I have to clear the table every day and my brother has to do the washing-up.
- **Ayer** tuve que **ir al supermercado con mi madre para hacer las compras**.
  Yesterday I had to go to the supermarket with my mother to do the shopping.
- **Mañana** tendré que **ordenar mi dormitorio. ¡Qué aburrido!**
  Tomorrow I will have to tidy my bedroom. How boring!

Talking About Pocket Money

ahorrar	to save
comprar	to buy
dar	to give
ganar	to earn
gastar	to spend
recibir	to receive
repartir periódicos	to deliver newspapers
trabajar en una tienda	to work in a shop
al mes	per month
por semana	per week
el dinero	money
una libra esterlina	one pound sterling

¿Te dan dinero tus padres? Do your parents give you pocket money?	**Sí, me dan veinte libras al mes.** Yes, they give me twenty pounds a month.
¿Tienes que ayudarles para ganar el dinero? Do you have to help them to earn your money?	**Sí, todos los días tengo que hacer algo.** Yes, I have to do something every day.
¿Qué haces? What do you do?	**Pues ayer tuve que pasar la aspiradora, hoy voy a hacer la comida y mañana tendré que fregar los platos.** Well, yesterday I had to do the vacuuming, today I'm going to do the cooking and tomorrow I will have to do the washing-up.
¿En qué gastas el dinero? What do you spend your money on?	**Lo gasto en ropa, revistas y caramelos y una vez al mes me gusta ir al cine con mis amigos. A veces, lo ahorro para comprar algo más caro.** I spend it on clothes, magazines and sweets and once a month I like to go to the cinema with my friends. Sometimes I save it to buy something more expensive.

Test Yourself

What do these mean in English?

1. **Nunca lavo el coche.**
2. **Mi hermana no hace nunca la comida.**

How do you say these in Spanish?

3. I wash the clothes.
4. She walks the dog.

Stretch Yourself

Say or write these in Spanish:

1. Today I have to do the washing-up and my brother has to tidy his bedroom.
2. Yesterday I had to walk the dog.

Practice Questions

Complete these exam-style questions to test your skills and understanding. Check your answers on page 91. You may wish to answer these questions on a separate piece of paper.

Reading

1. Choose one of the following adjectives to describe each of the people below:

simpático divertido perezoso listo deportivo reservado

a) Mi hermano menor no hace nada en casa y nunca quiere hacer los deberes.

.... (1)

b) Mi padre me ayuda mucho y es muy comprensivo.

.... (1)

c) Mi hermano mayor tiene un trabajo muy importante y tiene que saber mucho.

.... (1)

d) Tengo un amigo que no habla mucho en el colegio y nunca quiere salir con nosotros.

.... (1)

e) A mi me encanta jugar al rugby y también voy mucho al gimnasio.

.... (1)

f) Mi primo Miguel es muy extrovertido y gracioso. Me encanta salir con él.

.... (1)

2. What opinion do these students have about helping with household chores?

Mis hermanos y yo tenemos tareas distintas que tenemos que hacer pero ellos no hacen tanto como yo y eso no me gusta. Creo que es importante que todos hagamos algo para ayudar.	Laura
Tengo que planchar mi ropa pero lo hago en mi dormitorio mientras escucho música y el tiempo pasa rápidamente. Mi madre dice que planchar es aburrido – yo me divierto.	Manolo
Para mí, los quehaceres son una pérdida de tiempo. Prefiero pasar mi tiempo hablando por teléfono con mis amigos o viendo la televisión en vez de fregar los platos o pasar la aspiradora.	Nuria

By each of the names below, write P (positive), N (negative) or P+N (positive + negative).

a) Laura .... (1)

b) Manolo .... (1)

c) Nuria .... (1)

Speaking

3 You are discussing your home life via internet phone with a student in a partner school in Spain. Prepare to talk in Spanish about each of the points below.

a) Your family (family members / description)

....

b) Relationships within the family

....

c) Your house

....

d) Your daily routine

....

e) What you did yesterday at home and your opinion

....

f) How you help at home

....

g) Your thoughts on your home life

.... (10)

Writing

4 You are a celebrity and have been asked to write a short magazine article about yourself.

Write the article in Spanish and include information on the following:

- Personal information
- Your daily routine at home
- Your house now compared with your house before you were famous
- Who is the most important influence in your life and why
- How your home life is different now you are a celebrity
- Your ambitions for the future

....

....

....

....

.... (15)

How well did you do?

1–12	Try again	13–21	Getting there	22–27	Good work	28–34	Excellent!

Local Area

Describing Your Local Area

la región	region
el barrio	neighbourhood / district
una ciudad	city / town
un pueblo	small town / village
una aldea	village
en las afueras	in the outskirts
en el centro	in the centre
en el campo	in the countryside
en la costa	on the coast
un ayuntamiento	town hall
un banco	bank
un castillo	castle
un centro comercial	shopping centre
un cine	cinema
un habitante	inhabitant
un mercado	market
un parque infantil	children's playground
un polideportivo	sports centre
un supermercado	supermarket
una biblioteca	library
una bolera	bowling alley
una catedral	cathedral
una comisaría	police station
una fábrica	factory
una iglesia	church
una mezquita	mosque
una piscina	swimming pool
una tienda	shop
una urbanización	housing estate
correos	post office

Useful Adjectives for Describing a Place

aislado / a	isolated
antiguo / a	old
bonito /a	pretty
comercial	commercial
congestionado / a	congested
contaminado / a	polluted
feo / a	ugly / unsightly
grande	big
histórico / a	historic
industrial	industrial
limpio / a	clean
moderno / a	modern
pequeño / a	small
precioso / a	lovely / beautiful
residencial	residential
ruidoso / a	noisy
sucio / a	dirty
tranquilo / a	peaceful
pintoresco / a	picturesque
turístico / a	attractive to tourists

Town or Country?

Vivo en un pueblo pequeño en el campo.
I live in a small town in the countryside.

Es muy bonito pero a veces puede ser bastante aislado.
It's very pretty but at times it can be quite isolated.

Por un lado, me gusta la ciudad porque el transporte público es muy bueno y siempre hay algo que hacer.
On one hand, I like the city because the public transport is very good and there is always something to do.

Por otro lado, hay demasiado tráfico y puede ser muy ruidoso.
On the other hand, there is too much traffic and it can be very noisy.

Prefiero vivir en el campo porque es siempre muy tranquilo. Sin embargo, no hay mucho empleo.
I prefer to live in the countryside because it's always very peaceful. However, there isn't much work.

Talking About What You Can Do

To say what you / one can or cannot do in your town or area, use **se puede** + infinitive. For example:

En mi pueblo se puede cenar en un restaurante.
In my town you can have dinner in a restaurant.

En el campo se puede dar un paseo o montar a caballo.
In the countryside you can go for a walk or go horse-riding.

En la ciudad se puede viajar facilmente en metro.
In the city one can easily travel by underground train.

Vivo en un barrio tranquilo en las afueras de una ciudad. Para mí lo mejor es que no estoy muy lejos del centro de la ciudad y por eso, los sábados suelo ir de compras con mis amigos o vamos al cine o a la bolera. En el pasado no había mucho para los jóvenes pero ahora se puede hacer muchas cosas divertidas. En cuanto a los turistas, lo interesante es el castillo histórico al lado del río. Diría que la catedral también merece la pena visitar porque es preciosa.

I live in a peaceful neighbourhood in the outskirts of a city. For me, the best thing is that I'm not very far from the centre of the city and therefore, on Saturdays, I usually go shopping with my friends or we go to the cinema or to the bowling alley. In the past there wasn't much for young people to do but now you can do lots of fun things. As for the tourists, the interesting thing is the historic castle next to the river. I would also say that the cathedral is worth visiting because it is beautiful.

✓ Maximise Your Marks

When writing or speaking, especially if you are aiming for an A or A*, try to use a range of complex structures. So if you are describing your neighbourhood or town, do not just list the facilities – think of other ways of expressing yourself!

Build Your Skills: Using 'Lo' + Adjective

When we want to talk about the good or bad 'thing', we use the neuter form of the definite article, **lo**, instead of the masculine or feminine.

Lo + adjective means 'the...thing'. For example:

lo bueno	the good thing
lo malo	the bad thing
lo mejor	the best thing
lo peor	the worst thing
lo interesante	the interesting thing
lo raro	the strange thing
lo único	the only thing

This is a useful and slightly different way of expressing an opinion about something.

Lo mejor de mi pueblo es que está en la costa.
The best thing about my town is that it is on the coast.

Lo malo es la contaminación.
The bad thing is the pollution.

En el pasado, lo bueno era que no había tanto tráfico.
In the past, the good thing was that there wasn't as much traffic.

? Test Yourself

What do these mean in English?

1. **Vivo en una ciudad industrial.**
2. **Mi pueblo es muy ruidoso.**

How do you say these in Spanish?

3. My town can be quite isolated.
4. One can visit the cathedral.

★ Stretch Yourself

Say or write these in Spanish:

1. The best thing about my town is that you can do lots of fun things with your friends.
2. The good thing about my town is the new shopping centre.

Getting Around

Talking About Means of Transport

el medio de transporte	means of transport
el autobús	bus
el autocar	coach
el avión	plane
el barco	boat
la bicicleta / bici	bicycle / bike
el camión	lorry
el coche	car
el ferry	ferry
el helicóptero	helicopter
el metro	underground train
la moto	motorbike
el taxi	taxi
el tranvía	tram
el tren	train

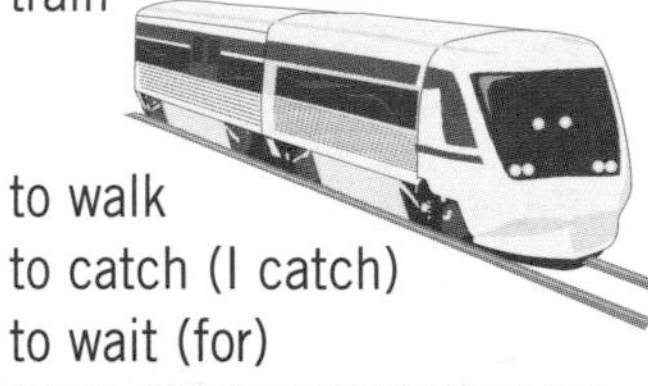

andar	to walk
coger (cojo)	to catch (I catch)
esperar	to wait (for)

✓ Maximise Your Marks

When saying how you travel, use **en** + transport, for example, **en coche** (by car). You do not need the definite article **el** / **la**. Note that 'on foot' is **a pie**.

El viaje es...	The journey is...
barato	cheap
caro	expensive
(in)cómodo	(un)comfortable
ecológico	ecological
económico	economical
lento	slow
limpio	clean
rápido	fast
sucio	dirty

Prefiero ir en coche porque es cómodo.
I prefer to go by car because it's comfortable.

No me gusta ir en coche a causa de los atascos. ¡Siempre hay retrasos!
I don't like to go by car because of the traffic jams. There are always delays!

Intento ir a pie a menudo porque hay que pensar en el medio ambiente.
I often try to go on foot because one has to think about the environment.

Imperatives

Imperatives are instructions or commands. We use them for giving directions, for example. In Spanish, the endings change depending on whether you are addressing the person(s) as **tú / vosotros** (familiar) or as **usted / ustedes** (formal).

Here are the imperative forms for the three main groups of verbs:

	Familiar		Formal		English
	tú (sing.)	vosotros (plural)	usted (sing.)	ustedes (plural)	
–ar	**Baja**	**Bajad**	**Baje**	**Bajen**	Go down!
–er	**Come**	**Comed**	**Coma**	**Coman**	Eat!
–ir	**Sube**	**Subid**	**Suba**	**Suban**	Go up!

The verbs below have irregular imperatives in the **tú** form:

decir → **di** (say)
hacer → **haz** (do)
ir → **ve** (go)
oír → **oye** (hear)
poner → **pon** (put)
salir → **sal** (go / get out)
tener → **ten** (have)
venir → **ven** (come)

¡Ven aquí! Come here!
¡Haz lo que te digo! Do what I say!

The **usted / ustedes** forms of the imperative use the subjunctive forms of the verb (see page 76). To form negative imperatives, use **No** followed by the subjunctive:

¡No hables! Don't speak! (**tú**)
¡No cruce aquí! Don't cross here! (**usted**)

Home and Away

Useful Phrases and Vocabulary

¿Dónde está..?	Where is...?	**la calle**	street
¿Para ir a...?	To get to...?	**la carretera**	main road
¿Por dónde se va a...?	How do I get to...?	**el cruce**	crossroads
Baje...	Go down...	**la esquina**	corner
Cruce...	Cross...	**la glorieta**	roundabout
Está a cien metros	It's 100 metres away	**el paso de peatones**	zebra crossing
Está a la derecha /	It's on the right	**la plaza mayor**	main square
Está a la izquierda	It's on the left	**el puente**	bridge
Tome la primera /	Take the first /	**el puerto**	port
segunda /	second /	**el río**	river
tercera calle	third street	**los semáforos**	traffic lights
a la derecha	on the right	**la señal**	sign
a la izquierda	on the left	**la zona peatonal**	pedestrianised area
Siga todo recto	Go straight on	**cerca de**	near to
Suba...	Go up...	**lejos de**	far from
Tuerza...	Turn...	**hasta**	as far as / up to

Asking For and Giving Directions

Note the use of **usted**. Conversations about directions often occur between strangers and are therefore formal.

¡Perdone! ¿Por dónde se va al polideportivo, por favor? → **A ver, siga todo recto, cruce la plaza y está a la izquierda.**
Excuse me! How do I get to the sports centre please? → Let's see, go straight on, cross the square and it's on the left.

¿Está lejos la catedral? → **No, está muy cerca. Suba la calle hasta los semáforos y tuerza a la derecha. Está a cincuenta metros.**
Is the cathedral far? → No, it's very close. Go up the street to the traffic lights and turn right. It's 50 metres away.

Build Your Skills: The Imperative of Reflexive Verbs

With reflexive verbs in the imperative, the reflexive pronoun is joined to the end of the verb. For an imperative using **vosotros**, the final **d** is dropped before adding the pronoun **os**:

¡Sentaos! Sit down!

The other forms all need an accent added to the stressed syllable:

¡Siéntate! Sit down! (**tú**)
¡Siéntese! Sit down! (**usted**)
¡Siéntense! Sit down! (**ustedes**)

For reflexive verbs in the negative imperative, the pronoun is not joined to the end:

¡No te sientes! Don't sit down!

? Test Yourself

What do these mean in English?

1. **Siga todo recto.**
2. **Cruce la carretera.**

How do you say these in Spanish?

3. It is on the left.
4. Go straight on as far as the traffic lights.

★ Stretch Yourself

Say or write these in Spanish:

1. Please get up early tomorrow! (**tú**)
2. Don't go down the main road! (**vosotros**)

Holiday Plans and Preparations

Countries and Continents

Alemania (f.)	Germany
Bélgica (f.)	Belgium
Dinamarca (f.)	Denmark
Escocia (f.)	Scotland
España (f.)	Spain
Estados Unidos (m. pl.)	United States
Francia (f.)	France
Gran Bretaña (f.)	Great Britain
Grecia (f.)	Greece
Inglaterra (f.)	England
Irlanda (f.)	Ireland
Irlanda del Norte (f.)	Northern Ireland
Islas Canarias (f. pl.)	Canary Islands
Italia (f.)	Italy
Holanda (f.)	Netherlands
Méjico / México (m.)	Mexico
(País de) Gales (m.)	Wales
Reino Unido (m.)	United Kingdom
Rusia (f.)	Russia
Suiza (f.)	Switzerland

los continentes	continents
África (f.)	Africa
América del Norte (f.)	North America
América del Sur (f.)	South America
Asia (f.)	Asia
Australia (f.)	Australia
Europa (f.)	Europe

¿De dónde eres?	Where are you from?
Soy de Escocia.	I am from Scotland.

Suelo ir de vacaciones a España porque mis padres tienen una casa allí.
I usually go on holiday to Spain because my parents have a house there.

Este año he decidido ir de vacaciones con mis amigos a Italia.
This year I have decided to go on holiday with my friends to Italy.

Languages

The language spoken in the country is usually the same as the masculine nationality – in Spain they speak **español**, in Germany they speak **alemán**.

Hablo francés e italiano.
I speak French and Italian.

Nationalities

Make sure you know the difference between the country and the nationality. You do not want to say that you went on holiday to Spanish (**español**), instead of to Spain (**España**)! This is a common mistake to make. Note that nationalities in Spanish do not have a capital letter.

Soy... / Es...	I am.../ He / She is...
alemán / alemana	German
argentino / a	Argentinian
británico / a	British
chileno / a	Chilean
cubano / a	Cuban
colombiano / a	Columbian
escocés / escocesa	Scottish
español / española	Spanish
francés / francesa	French
galés / galesa	Welsh
holandés / holandesa	Dutch
inglés / inglesa	English
italiano / a	Italian
(nor)irlandés / (nor)irlandesa	(Northern) Irish
norteamericano / a	North American
peruano / a	Peruvian

Soy galés.	**María es española.**
I am Welsh (male).	María is Spanish.

Useful Verbs

bañarse	to swim / bathe
broncearse	to get a suntan
dar un paseo	to go for a walk
descansar	to rest
esquiar	to ski
hacer turismo	to go sightseeing
hacer la maleta	to pack your suitcase
hacer una visita guiada	to go on a guided tour
ir de excursión	to go on an excursion
nadar	to swim
pasar (una quincena)	to spend (a fortnight)
practicar los deportes acuáticos / de invierno	to practise water sports / winter sports
quedarse	to stay
sacar fotos	to take photos
tomar el sol	to sunbathe
viajar	to travel
volar	to fly

More Time Expressions

el año pasado	last year
la semana pasada	last week
cuando era pequeño / a	when I was little
hace dos años	two years ago
cada año	every / each year
en agosto	in August
en invierno	in winter
en otoño	in autumn
en primavera	in spring
en verano	in summer
este verano	this summer
el año próximo / que viene	next year
el mes próximo	next month
mañana	tomorrow
pasado mañana	the day after tomorrow
ya	already
otra vez	again

✓ Maximise Your Marks

When talking about holiday dates, use **desde** to mean 'from' and **hasta** to mean 'until'.
For example:

- **Voy a pasar una semana en España desde el tres hasta el diez de mayo.**
  I am going to spend a week in Spain from the 3rd to the 10th of May.

You can also use **de** and **a**:

- **Voy a pasar una semana en España del tres al diez de mayo.**

The Immediate Future

There are two future tenses in Spanish (see also page 73). The immediate future describes what *is going to* happen. It uses the verb **ir** (to go) in the present tense + **a** + infinitive.

Ir	To go		
Voy	I go / am going		**viajar**
Vas	You go / are going (**tú**)		**visitar**
Va	He / She / It goes / is going;		**esquiar**
	You go / are going (**Vd.**)	**a**	**nadar**
Vamos	We go / are going		**hablar**
Vais	You go / are going (**vosotros**)		**comer**
Van	They go / are going;		**salir**
	You go / are going (**Vds.**)		

Este año, voy a ir a España con mi familia.
This year, I am going to go to Spain with my family.

El año próximo vamos a visitar Alemania y Francia.
Next year we are going to visit Germany and France.

Build Your Skills: Using 'Antes de' and 'Después de'

Vary your sentence structure by using **antes de** + infinitive to say 'before doing something' or **después de** + infinitive to say 'after doing something'.
For example:

¡Voy a hacer la maleta una semana antes de irme para no olvidarme de nada!
I am going to pack my suitcase a week before I go so that I don't forget anything!

Después de esquiar en las montañas, voy a pasar una semana en la playa.
After skiing in the mountains, I'm going to spend a week at the beach.

? Test Yourself

What do these mean in English?

1. **Normalmente voy de vacaciones a Francia.**
2. **Voy a broncearme en la playa.**

How do you say these in Spanish?

3. Next year I am going to go to Spain.
4. We are going to take lots of photos.

★ Stretch Yourself

Say or write these in Spanish:

1. Before going sightseeing, I am going to swim in the pool and sunbathe a little.
2. After eating in the restaurant, I'm going to go for a walk.

Accommodation

Talking About Accommodation

el aire acondicionado	air conditioning	**la habitación doble**	double room
el albergue juvenil	youth hostel	**el huésped**	guest
el alojamiento	accommodation	**la llave**	key
el aparcamiento	car park	**la media pensión**	half board
el apartamento	apartment	**la pensión**	bed and breakfast
el ascensor	lift	**la pensión completa**	full board
el aseo	toilet	**la posada**	inn / hostel
el balcón	balcony	**la vista al mar**	view of the sea
la cuenta	bill	**alojarse**	to stay / lodge
el hotel	hotel	**alquilar**	to hire / rent
la habitación individual	single room	**reservar**	to reserve

Forming a Question

There are two ways of forming a question in Spanish. When speaking, the easiest is to take a straightforward sentence and then use intonation to turn it into a question – by raising your voice at the end. In written Spanish, you do this by adding question marks at the beginning and end. Hence **La habitación tiene balcón** (The room has a balcony) becomes **¿La habitación tiene balcón?** (Does the room have a balcony?)

Here are some more examples:

Tiene habitaciones libres. You have vacant rooms.	→	**¿Tiene habitaciones libres?** Do you have any vacant rooms?
Te gusta la vista. You like the view.	→	**¿Te gusta la vista?** Do you like the view?

The other way to form a question is by using a question word or interrogative pronoun. Here are the most common question words:

¿Adónde?	Where...to?	→	**¿Adónde vas mañana?**	Where are you going (to) tomorrow?
¿Cómo?	What / How?	→	**¿Cómo te llamas?**	What are you called?
¿Cuál(es)?	What / Which (one)?	→	**¿Cuál quieres?**	Which (one) do you want?
¿Cuándo?	When?	→	**¿Cuándo te lo dijo?**	When did he tell you?
¿Cuánto?	How much / long?	→	**¿Cuánto es?**	How much is it?
¿Cuántos...?	How many...?	→	**¿Cuántas bicicletas hay?**	How many bikes are there?
¿Dónde?	Where?	→	**¿Dónde vives?**	Where do you live?
¿Por qué?	Why?	→	**¿Por qué no te gusta la lluvia?**	Why don't you like the rain?
¿Qué?	What?	→	**¿Qué has hecho?**	What have you done?
¿Quién?	Who?	→	**¿Quién es esa chica?**	Who's that girl?

Remember to put **¿** at the beginning of the question and **?** at the end. The same applies to exclamation marks (**¡** and **!**) Also, the question words all have an accent on the main vowel.

Organising Accommodation

Quisiera reservar una habitación doble con cuarto de baño y vistas al mar para dos noches.
I'd like to reserve a double room with a bathroom and sea views for two nights.

¿Cuánto es por noche por persona?
How much is it per night per person?

Quisiera quedarme una semana desde el tres hasta el diez de junio.
I'd like to stay for a week from the 3rd to the 10th of June.

¿A qué hora se sirve el desayuno?
What time is breakfast served?

Build Your Skills: Talking About Problems

Even if your holiday goes without any hitches, it makes your speaking and writing more interesting if you can talk about a problem with the accommodation, etc. Just make it up! Here are some examples:

No hay agua caliente en el bloque sanitario.
There isn't any hot water in the toilet block.

El ascensor / el aseo / la ducha / la luz no funciona.
The lift / toilet / shower / light doesn't work.

He perdido mi bolso / la llave / el monedero / el pasaporte.
I have lost my handbag / key / purse / passport.

No hay papel higiénico / jabón / toallas.
There isn't / aren't any toilet paper / soap / towels.

La habitación y la cama están sucias.
The room and the bed are dirty.

✓ Maximise Your Marks

Add more depth to your work by giving specific detail rather than just a basic description.
For example:

- **Nos quedamos en un camping muy bonito al lado de un río con vistas preciosas del campo. Había un bloque sanitario y un parque infantil. Pero no había ni agua caliente ni papel higiénico. ¡Qué horror!**
  We stayed on a very pretty campsite by a river with beautiful views of the countryside. There was a toilet block and a children's play area. But there was no hot water or toilet paper! How awful!

At the Campsite

el agua caliente	hot water
el agua potable	drinking water
el bloque sanitario	toilet block
el camping	campsite
la caravana	caravan
el saco de dormir	sleeping bag
el sitio	a place
la tienda (de campaña)	tent
montar la tienda	to pitch the tent

¿Hay sitio para una tienda para tres noches?
Is there space for a tent for three nights?

Somos dos adultos y dos niños y tenemos un coche y una caravana.
There are two adults and two children and we have a car and a caravan.

¿Hay agua potable aquí?
Is there drinking water here?

? Test Yourself

What do these mean in English?

1. **Quisiera una habitación individual.**
2. **Somos cuatro adultos y tres niños.**

How do you say these in Spanish?

3. I'd like a double room, please.
4. Is there space for a caravan and a car?

★ Stretch Yourself

Say or write these in Spanish:

1. The light in my room doesn't work and there aren't any towels in the bathroom.
2. I have lost my keys and my purse.

Home and Away

The Preterite Tense

Using the Preterite

The preterite tense is also known as the simple past. It is used to talk about completed actions in the past – 'I went', 'You ate', 'He listened' are examples in English. Using the preterite is very important in your GCSE. You need to use it to talk about past events and it is one of the key factors to help you to achieve a grade C or above.

There are two sets of endings for regular verbs in the preterite: one for **–ar** verbs and the other for **–er** and **–ir** verbs. Here they are in full:

The Preterite of '–ar' Verbs

Hablar	To speak
Hablé	I spoke
Hablaste	You spoke (**tú**)
Habló	He / She / It spoke; You spoke (**Vd.**)
Hablamos	We spoke
Hablasteis	You spoke (**vosotros**)
Hablaron	They spoke; You spoke (**Vds.**)

Hablé español en México.
I spoke Spanish in Mexico.

Hablaron por teléfono.
They spoke on the phone.

The Preterite of '–er' and '–ir' Verbs

Comer	To eat
Comí	I ate
Comiste	You ate (**tú**)
Comió	He / She / It ate; You ate (**Vd.**)
Comimos	We ate
Comisteis	You ate (**vosotros**)
Comieron	They ate; You ate (**Vds.**)

Vivir	To live
Viví	I lived
Viviste	You lived (**tú**)
Vivió	He / She / It lived; You lived (**Vd.**)
Vivimos	We lived
Vivisteis	You lived (**vosotros**)
Vivieron	They lived; You lived (**Vds.**)

¿Comiste mucho pescado?
Did you eat lots of fish?

Comí platos típicos en España.
I ate traditional dishes in Spain.

Vivimos en un piso en el centro.
We lived in a flat in the centre of town.

Radical-changing Verbs in the Preterite

There are no **–ar** or **–er** radical-changing verbs in the preterite tense. For **–ir** verbs, there are two types of changes: **e** > **i** and **o** > **u** in the third persons singular (**él**, **ella**, **usted**) and plural (**ellos**, **ellas**, **ustedes**):

The 'e' > 'i' Change

Pedir	To ask for
Pedí	I asked for
Pediste	You asked for (**tú**)
Pidió	He / She / It asked for; You asked for (**Vd.**)
Pedimos	We asked for
Pedisteis	You asked for (**vosotros**)
Pidieron	They asked for; You asked for (**Vds.**)

Some other **–ir** verbs that follow this pattern are:

preferir	to prefer
reir	to laugh
seguir	to follow
sentir	to feel
sonreir	to smile
vestirse	to get dressed

Prefirió viajar en avión.
He preferred to travel by plane.

Siguieron las direcciones.
They followed the directions.

The 'o' > 'u' Change

Dormir	To sleep
Dormí	I slept
Dormiste	You slept (**tú**)
Durmió	He / She / It slept; You slept (**Vd.**)
Dormimos	We slept
Dormisteis	You slept (**vosotros**)
Durmieron	They slept; You slept (**Vds.**)

The reflexive verb **morir(se)** (to die) also follows this pattern.

✓ Maximise Your Marks

In order to achieve the higher grades you must demonstrate a sound knowledge of verbs in the preterite tense. When speaking or writing, try to add in some irregular verb forms as well as the regular ones to 'show off' to the examiners. This will help you to achieve more marks.

Irregular Verbs in the Preterite

The most common irregular verbs in the preterite are as follows:

	Ser / Ir To be / To go	Hacer To do / To make	Tener To have	Ver To see
(yo)	fui	hice	tuve	vi
(tú)	fuiste	hiciste	tuviste	viste
(él / ella / usted)	fue	hizo	tuvo	vio
(nosotros)	fuimos	hicimos	tuvimos	vimos
(vosotros)	fuisteis	hicisteis	tuvisteis	visteis
(ellos / ellas / ustedes)	fueron	hicieron	tuvieron	vieron

	Dar To give	Poner To put	Poder To be able	Venir To come
(yo)	di	puse	pude	vine
(tú)	diste	pusiste	pudiste	viniste
(él / ella / usted)	dio	puso	pudo	vino
(nosotros)	dimos	pusimos	pudimos	vinimos
(vosotros)	disteis	pusisteis	pudisteis	vinisteis
(ellos / ellas / ustedes)	dieron	pusieron	pudieron	vinieron

Note that none of these has an accent.

Some verbs have an irregular spelling only in the first person singular (**yo**) of the preterite:

buscar	to look for	→ **busqué**	I looked for
sacar	to take out / get	→ **saqué**	I took out / got
tocar	to play (instrument)	→ **toqué**	I played
cruzar	to cross	→ **crucé**	I crossed
empezar	to start	→ **empecé**	I started
jugar	to play	→ **jugué**	I played
llegar	to arrive	→ **llegué**	I arrived

Llegué muy tarde al hotel.
I arrived very late at the hotel.

¿Visteis la plaza mayor en el centro de la ciudad?
Did you see the main square in the centre of the city?

The verbs **leer** (to read) and **caer** (to fall) follow their own pattern:

Leí	I read
Leíste	You read (**tú**)
Leyó	He / She / It read; You read (**Vd.**)
Leímos	We read
Leísteis	You read (**vosotros**)
Leyeron	They read; You read (**Vds.**)
Caí	I fell
Caíste	You fell (**tú**)
Cayó	He / She / It fell; You fell (**Vd.**)
Caímos	We fell
Caísteis	You fell (**vosotros**)
Cayeron	They fell; You fell (**Vds.**)

Other verbs that follow this pattern are:

construir	to build	**destruir**	to destroy
creer	to believe	**oír**	to hear

Test Yourself

What do these mean in English?

1. **Bebieron una botella de agua.**
2. **Salí anoche con mis amigos.**

How do you say these in Spanish?

3. I arrived   I visited   I saw
4. You (**tú**) ate   We drank   You (**tú**) slept

Stretch Yourself

Say or write these in Spanish:

1. Last year my dad read the menu in French.
2. I took lots of photos of the hotel.

Holiday Activities

Talking About Holiday Activities

bailar en la discoteca	to dance at the disco
cenar en un restaurante	to eat in a restaurant
comprar recuerdos	to buy souvenirs
dar una vuelta en bici	to go for a bike ride
durar	to last
mejorar mi español	to improve my Spanish
montar a caballo	to go horse-riding
pasar (tiempo)	to spend (time)
practicar el idioma	to practise the language
relajarse	to relax
ver lugares de interés	to see interesting places
visitar	to visit
volver	to return

Suelo ir de vacaciones con mi familia y a veces viene un amigo también.
I usually go on holiday with my family and sometimes a friend comes as well.

Normalmente visito muchos sitios de interés porque me gusta conocer culturas diferentes.
I usually visit lots of interesting places because I like to find out about different cultures.

Por la noche, a veces cenamos en un restaurante y comemos platos típicos y otras veces me gusta dar una vuelta por el pueblo y hablar con gente para practicar el idioma.
In the evening, sometimes we dine in a restaurant and eat traditional food and other times I like to go for a walk in the town and speak to people to practise the language.

El año pasado fui a Kenya con mi familia. Viajamos en avión y el vuelo duró ocho horas. ¡Fue muy emocionante!
Last year I went to Kenya with my family. We travelled by plane and the flight lasted eight hours. It was really exciting!

Visitamos muchos lugares bonitos y un día fuimos de safari.
We visited lots of lovely places and one day we went on safari.

Vimos a muchos animales impresionantes y lo pasé fenomenal. Me encantaría volver algún día.
We saw lots of amazing animals and I had a brilliant time. I'd love to go back some day.

The Weather in the Past

To talk about the weather on a specific day or occasion in the past, use the preterite tense:

El domingo hubo niebla.
On Sunday it was foggy.

Un día hizo calor.
One day it was hot.

To talk about the weather in the past over a period of time, use the imperfect tense:

¿Qué tiempo hacía?
What was the weather like?

Hacía sol / frío / calor / buen tiempo / mal tiempo.
It was sunny / cold / hot / fine weather / bad weather.

Estaba nublado / lloviendo / nevando.
It was cloudy / raining / snowing.

Había niebla.
It was foggy.

Conjunctions or Connectives

Try to use a variety of conjunctions or connectives (linking words) to extend your sentences rather than using short phrases all of the time. This will help you to gain more marks. Here are the most useful:

a pesar de	in spite of
además	besides
así que	so that / therefore
aunque	although
como	as / like
cuando	when
o / u	or
pero	but
por eso	therefore
porque	because
pues	then
salvo que	except
si	if
sin embargo	however
tal vez	maybe
también	also
y / e	and
ya que	since

Fuimos a la playa como a mí me gusta tanto.
We went to the beach as I like it so much.

No visitamos un museo porque es aburrido.
We didn't visit a museum because it's boring.

'And' is **y** unless it is followed by a word beginning with **i** or **hi**, in which case it changes to **e**:

- **Aprendo español e inglés.**
  I study Spanish and English.

'Or' is **o** unless it is followed by a word beginning with **o** or **ho**, in which case it changes to **u**:

- **Normalmente desayuno a las siete u ocho.**
  I usually have breakfast at seven or eight.

Build Your Skills: Using Sequencers for Structure

A good way to show the order of events in what you want to say or write is to use sequencers. These are words such as **primero...** (first); **antes (de)...** (before); **después (de)...** (afterwards); **luego...** (then); **entonces...** (then / so); **mientras** (while / meanwhile); **tan pronto como** (as soon as); **finalmente / por fin...** (finally), etc. Sequencers help to structure your work and make it 'flow' better.

Primero, visitamos Buenos Aires. Dimos un paseo por el centro y luego comimos en un restaurante típico. Después visitamos un museo.

First of all, we visited Buenos Aires. We went for a walk around the city centre and then we ate in a traditional restaurant. Afterwards we visited a museum.

Boost Your Memory

Make a list of some useful phrases, such as conjunctions and sequencers, that can be used over and over again in all topics. Have the list somewhere handy so that you can refer to it all of the time – you will start to remember the phrases much more easily.

Test Yourself

What do these mean in English?

1. **Siempre bailo en la discoteca.**
2. **Hacía buen tiempo – sol y mucho calor.**

How do you say these in Spanish?

3. I saw lots of interesting places.
4. I am going to return next summer.

Stretch Yourself

Say or write this in Spanish:

1. First we went to the beach while my parents relaxed on the balcony. Then we visited a pretty town and finally I ate in a restaurant with my family and friends.

Practice Questions

Complete these exam-style questions to test your skills and understanding. Check your answers on page 92. You may wish to answer these questions on a separate piece of paper.

Reading

1. Read the sentences below and match them to the statements that follow.

A **Vivo en un pueblo pequeño en el campo que es bonito, pero a veces puede ser bastante aislado.**

B **Para mí es muy residencial y no hay muchas instalaciones.**

C **Es muy fácil divertirse en mi pueblo a causa de las instalaciones que hay.**

D **Lo que ofrece es la tranquilidad y un paisaje precioso.**

E **Lo que pasa es que hay basura en las calles y contaminación de las fabricas.**

F **El castillo pintoresco y la catedral antigua merecen la pena visitar.**

a) Lots to do .... (1)

b) Historic sites .... (1)

c) Pollution .... (1)

d) Town is isolated .... (1)

e) Positives of the countryside .... (1)

f) Not a lot to do there .... (1)

2. Read the following passage about Rosa's holiday. Then answer the questions that follow in English.

En julio del año pasado pasé un mes con mi familia en la costa en nuestro apartamento. Me encanta estar allí porque puedo ver a mis amigos y paso todos los días en la playa con ellos. Por la mañana solemos bañarnos en el mar y a mí me gusta mucho practicar los deportes acuáticos. El año pasado fui un día con mi familia a un pueblo en la montaña y dimos una vuelta en bici, que me encantó. ¡Fue precioso! Luego saqué muchas fotos y compré unos recuerdos para mis abuelos. Por la noche volvimos a la costa y cenamos en un restaurante cerca del apartamento. ¡Fue un día perfecto!

a) How long did Rosa go on holiday for? .... (1)

b) Where did she stay? .... (1)

c) Why does she like going there? .... (1)

d) What does she usually do in the morning? .... (1)

e) Where did she go one day last year? .... (1)

f) What did she do after the bike ride? .... (1)

g) Where did they have dinner? .... (1)

Speaking

3 You are being interviewed for a school magazine about your local area by a Spanish student. Prepare to talk in Spanish about each of the points below.

a) Your neighbourhood and nearby leisure facilities

....

b) A description of your town / city

....

c) What you can do there

....

d) What it used to be like

....

e) Your opinion of your local area and why

....

f) How you travel around

....

g) Your thoughts on where you will live in the future

.... (10)

Writing

4 You are entering a competition to win a holiday and have to write a description of a past holiday of your own. Write the description in Spanish and include information on the following:

- Where you went and how you got there
- Where you stayed
- What you did while you were there
- The best day of your holiday and why
- What type of holiday you prefer and why
- Your future holiday plans

....

....

....

....

....

.... (15)

How well did you do?

1–13	Try again	14–23	Getting there	24–31	Good work	32–38	Excellent!

Leisure and Pastimes

Free-time Activities

bailar	to dance
cantar	to sing
charlar con amigos	to chat to friends
dar una vuelta	to go for a walk
divertirse	to enjoy oneself / have fun
escuchar música	to listen to music
ir a la bolera	to go bowling
ir al cine / al teatro	to go to the cinema / theatre
ir a una corrida de toros	to go to a bullfight
jugar al ajedrez	to play chess
jugar a las cartas	to play cards
jugar con los videojuegos	to play videogames
leer	to read
navegar por Internet	to surf the Internet
pasarlo bien / bomba	to have a good / great time
pintar	to paint
quedar con amigos	to meet up with friends
salir con amigos	to go out with friends
tocar un instrumento	to play an instrument
ver la televisión	to watch television
ver un partido	to watch a match / game

Watching TV

los anuncios	adverts
el canal	channel
los concursos	gameshows
los documentales	documentaries
las emisiones	programmes / broadcasts
las emisiones deportivas	sports programmes
las noticias / el telediario	(TV) news
la pantalla	screen
el programa	programme
los programas musicales	music programmes
las series	series
las telenovelas	soaps
la TVE	Spanish TV channel

¿Qué ponen en la tele esta noche?
What's on TV tonight?

Hay un concurso muy gracioso a las ocho. ¿Te apetece verlo?
There's a very funny gameshow on at eight o'clock. Do you fancy watching it?

Vale, sí. ¿Viste el documental anoche en la 2? Fue muy informativo e interesante.
OK, yes. Did you see the documentary last night on channel 2? It was really informative and interesting.

Mis programas preferidos son los dibujos animados.
My favourite programmes are cartoons.

At the Cinema

el cine	cinema
la comedia	comedy
los dibujos animados	cartoons
la película	film
la película romántica / de amor	romantic film
la película de acción / aventura	action / adventure film
la película de ciencia-ficción	science-fiction film
la película de guerra	war film
la película de terror	horror film
la película del oeste	western
la película policíaca	detective film
la taquilla	ticket office

A mí me gustan las películas de aventura y también las comedias. No me gustan las películas de ciencia-ficción porque me parecen aburridas y tontas.
I like adventure films and comedies too. I don't like science-fiction films because they seem boring and stupid to me.

Ayer fui al cine con mis amigos y vi una película de Harry Potter que me encantó.
Yesterday I went to the cinema with my friends and I saw a Harry Potter film, which I loved.

Se trata de un grupo de amigos que son magos en un colegio de magia.
It's about a group of friends who are wizards in a school of magic.

Build Your Skills: Comparatives and Superlatives

Instead of just giving an opinion about one thing, you can make your speaking and writing more interesting by comparing things.

To compare two things, we use comparatives. In English, these take the form 'funnier than', 'more interesting than', 'less entertaining than', for example. In Spanish, comparatives are formed using these words (plus an adjective where indicated '...'):

más...que	more...than
menos...que	less...than
mejor que	better than
peor que	worse than
tan...como	as...as

Do not forget that where necessary the adjectives must agree.

Me gustan los dibujos animados porque son más divertidos que los documentales.
I like cartoons because they are more amusing than documentaries.

Los programas musicales son mejores que las emisiones deportivas.
Music programmes are better than sports programmes.

To compare more than two things, to say that something or someone is 'the best', 'the worst', 'the most interesting', etc., we use the superlative. In Spanish, this is formed by putting **el**, **la**, **los** or **las** in front of **más**:

- **Las películas de terror son las más emocionantes.**
  Horror films are the most exciting.

Common irregular superlatives are:

el / la mejor	the best
el / la peor	the worst
el / la mayor	the biggest / oldest
el / la menor	the youngest

Another way of expresssing a superlative in Spanish is to add **–ísimo** to the end of the adjective, after removing the final vowel:

- **Las telenovelas son aburridísimas.**
  Soaps are extremely boring.
- **Mi actor preferido es guapísimo.**
  My favourite actor is extremely handsome.

✓ Maximise Your Marks

Showing that you can express yourself in many different ways is one of the key techniques that will gain you extra marks. A straightforward but effective way of expressing an opinion is to use **¡Qué** + adjective**!** Remember to add the exclamation marks in your writing; and, in your speaking, show some intonation in your voice.

¡Qué bien!	(How) great!
¡Qué emocionante!	How exciting!
¡Qué guay!	How cool!
¡Qué horror!	How awful!
¡Qué lío!	What a mess!
¡Qué rico!	How delicious!
¡Qué suerte!	What (good) luck! / How lucky!
¡Qué va!	No way!
¡Qué vergüenza!	How embarrassing!

? Test Yourself

What do these mean in English?

1. **Las telenovelas son mis programas preferidos.**
2. **Por la noche prefiero navegar por Internet.**

How do you say these in Spanish?

3. I love going to the cinema with my friends.
4. Last week I went to the theatre with my cousin.

★ Stretch Yourself

Say or write these in Spanish:

1. Documentaries are more interesting than the news.
2. Sports programmes are less boring than cartoons.

Sports and Exercise

Sports

el alpinismo	climbing
el atletismo	athletics
el baloncesto	basketball
el balonmano	handball
el béisbol	baseball
el billar	billiards
el boxeo	boxing
el ciclismo	cycling
los deportes	sports
los deportes de riesgo	extreme sports
la equitación	horse-riding
el esquí	skiing
el footing	jogging
el fútbol	football
la gimnasia	gymnastics
el monopatín	skateboarding
la natación	swimming
el patinaje (sobre hielo)	skating (ice)
la pesca	fishing
el ping-pong	table tennis
el rugby	rugby
el tenis	tennis
el tenis de mesa	table tennis
la vela	sailing
el voleibol	volleyball
el windsurf	windsurfing

The sports and activities listed are all used with the following verbs: **jugar a** (to play); **practicar** (to practise); **hacer** (to do). The general rule is that if it is a game, such as football or basketball, you use **jugar a**, and for the other activities you use **practicar** or **hacer**. With sports, **hacer** can be translated into English as to 'do / practise / go'. For example:

- **Practico / hago la vela los fines de semana.**
  I practise / do / go sailing on weekends.

Jugar is always followed by **a**, so take care if the sport is masculine (which the majority are), as you will need to contract the **a** + **el** to make **al**:

- **Juego al fútbol todos los días.**
  I play football every day.
- **Suelo jugar al baloncesto en el colegio.**
  I usually play basketball at school.

Other Useful Verbs

correr	to run
dar un paseo	to go for a walk
entrenarse	to train
esquiar	to ski
ganar la copa	to win the cup
marcar un gol	to score a goal
montar	to ride
nadar	to swim
patinar	to skate
ir de pesca	to go fishing
perder	to lose
saltar	to jump
ser hincha de	to be a fan of

Boost Your Memory

There is often more than one way of saying the same thing in Spanish (and in English!), particularly when talking about sports and activities. For example, to talk about swimming you could use any of the following:

nadar	to swim
practicar / hacer la natación	to go swimming
bañarse	to bathe / swim
ir a la piscina	to go to the pool

You need to learn different ways of saying the same thing, so always try to group like phrases together when you note them down. This will help you to recall them in the exam.

Sports People and Places

el / la atleta	athlete
el campeón	champion
la campeona	champion
el / la ciclista	cyclist
el equipo	team
el / la espectador / ora	spectator
el / la futbolista	footballer
el / la jugador / ora	player
el / la tenista	tennis player
el campo	pitch / field
la cancha (de tenis)	(tennis) court
el estadio	stadium
la piscina	swimming pool
la pista	ski-slope
la pista de hielo	ice-rink
el polideportivo	sports centre

Talking About Sports

Me encantan los deportes. Juego al hockey dos veces a la semana y también practico la natación y el footing.
I love sports. I play hockey twice a week and I also go swimming and jogging.

Soy hincha del atletismo y lo practico con mi equipo escolar cada semana.
I am an athletics fan and I practise it every week with my school team.

Soy miembro del club de tenis de mesa en mi colegio.
I am a member of the table-tennis club at my school.

Mi deporte preferido es la natación y entreno en la piscina todos los días.
My favourite sport is swimming and I train at the pool every day.

Cuando era más joven no me gustaban mucho los deportes, pero ahora me encanta practicar el esquí.
When I was younger I didn't really like sports, but now I love to go skiing.

En el futuro, me gustaría practicar un deporte de riesgo porque ¡es más emocionante!
In the future, I would like to do an extreme sport because it's more exciting!

Build Your Skills: Asking and Answering 'How Long?'

To say how long you *have been doing* something, you use **desde hace** with the present tense of the verb. This is different from English and is a very useful phrase, so make sure that you learn it!

¿Desde hace cuánto tiempo practicas la gimnasia?
How long have you been practising gymnastics?

Practico la gimnasia desde hace cinco años.
I've been doing gymnastics for five years.

If you use **desde hacía** with the imperfect tense, you can say how long you *had* been doing something:

Jugaba al béisbol desde hacía un año cuando ganamos la copa.
I had been playing baseball for a year when we won the cup.

Test Yourself

What do these mean in English?

1. **Practico el monopatín en el parque.**
2. **Juego al tenis de mesa todos los días en el colegio.**

How do you say these in Spanish?

3. I usually practise boxing at the weekend.
4. I'm going to go fishing with my brother on Saturday.

Stretch Yourself

Say or write these in Spanish:

1. I have been practising ice-skating for two years and I have to go to the ice-rink twice a week.
2. I had been doing athletics for five years.

More Free-time Activities

Build Your Skills: The Imperfect Tense

The imperfect tense is used in Spanish to describe:

- What *used to* happen in the past:
  Veíamos la tele todas las noches.
  We used to watch TV every night.
- what someone or something *was like*:
  La casa era muy grande.
  The house was very big.
- what *was happening* / what someone *was doing* (when something else happened):
  Escuchaba la radio cuando llegó mi abuela.
  I was listening to the radio when my grandmother arrived.

To form the imperfect, add these endings to the verb stem:

'–ar' Verbs

Hablar To speak
Hablaba I was speaking / used to speak
Hablabas You were speaking / used to speak (**tú**)
Hablaba He / She / It was speaking / used to speak; You were speaking / used to speak (**Vd.**)
Hablábamos We were speaking / used to speak
Hablabais You were speaking / used to speak (**vosotros**)
Hablaban They were speaking / used to speak; You were speaking / used to speak (**Vds.**)

'–er' Verbs

Comer To eat
Comía I was eating / used to eat
Comías You were eating / used to eat (**tú**)
Comía He / She / It was eating / used to eat; You were eating / used to eat (**Vd.**)
Comíamos We were eating / used to eat
Comíais You were eating / used to eat (**vosotros**)
Comían They were eating / used to eat; You were eating / used to eat (**Vds.**)

'–ir' Verbs

Vivir To live
Vivía I was living / used to live
Vivías You were living / used to live (**tú**)
Vivía He / She / It was living / used to live; You were living / used to live (**Vd.**)
Vivíamos We were living / used to live
Vivíais You were living / used to live (**vosotros**)
Vivían They were living / used to live; You were living / used to live (**Vds.**)

There are only three irregular verbs in the imperfect tense:

	Ir To go	**Ser** To be	**Ver** To see
(**yo**)	**iba**	**era**	**veía**
(**tú**)	**ibas**	**eras**	**veías**
(**él / ella / usted**)	**iba**	**era**	**veía**
(**nosotros**)	**íbamos**	**éramos**	**veíamos**
(**vosotros**)	**ibais**	**erais**	**veíais**
(**ellos / ellas / ustedes**)	**iban**	**eran**	**veían**

Two useful verbs in the imperfect are **haber** (**hay** in the present, meaning 'there is / there are') and **soler** (to used to):

- **Había mucha gente allí.**
  There were a lot of people there.
- **Solía salir con mis amigos los sábados.**
  I used to go out with my friends on Saturdays.

✓ Maximise Your Marks

By including a few complex sentences in your speaking and writing, you will really impress the examiner. To gain an A or A*, try to include the preterite and the imperfect in the same sentence to show that you know your tenses! For example:

- **Veía la película con mi novia cuando me llamó mi madre. ¡Qué vergüenza!**
  I was watching the film with my girlfriend when my mother rang. How embarrassing!

Music

Me encanta escuchar música sobre todo cuando hago mis deberes.
I love listening to music especially when I'm doing my homework.

Lo que más me gusta es la música pop pero no soporto la música hip-hop.
What I like the most is pop music but I can't stand hip-hop.

Siempre llevo mi iPod conmigo a todos los sitios y así puedo escuchar música cuando quiera.
I always take my iPod everywhere with me and then I can listen to music whenever I want.

Mi cantante preferido es...porque tiene una voz preciosa y también ¡es muy guapo!
My favourite singer is...because he has a lovely voice and he's also really good-looking!

la batería	drums
las castañuelas	castanets
la flauta	flute
la guitarra	guitar
el piano	piano
la trompeta	trumpet
el violín	violin

You already know that the verb **jugar a** (to play) is used with sports. If you want to say you play a musical instrument, however, the verb is **tocar**.

¿Sabes tocar un instrumento?
Can you play an instrument?

Sé tocar el piano y la guitarra.
I can play the piano and the guitar.

Reading

la lectura	reading
el libro	book
la novela	novel
el periódico / el diario	newspaper
la prensa	the press
la revista	magazine
el tebeo	comic

Me encanta la lectura. Paso mucho tiempo leyendo novelas y revistas.
I love reading. I spend a lot of time reading novels and magazines.

Yo no leo mucho porque lo encuentro muy aburrido.
I don't read a lot because I find it very boring.

Leo el periódico todos los días porque creo que es importante estar al día con las noticias.
I read the newspaper every day because I think it is important to be up-to-date with the news.

Test Yourself

What do these mean in English?

1. **No sé tocar un instrumento pero me encanta escuchar música.**
2. **Siempre leo revistas sobre la moda porque me interesa la ropa.**

How do you say these in Spanish?

3. I watch the TV to relax.
4. In my opinion (for me) it's really boring.

Stretch Yourself

Say or write these in Spanish:

1. We used to go to the swimming pool on Saturday mornings and I used to swim with my brothers.
2. When I was young, I used to pay the trumpet.

Clothes and Fashion

Clothes

la ropa	clothes
un abrigo	coat
un bañador	swimsuit
una blusa	blouse
unas botas	boots
unos calcetines	socks
una camisa	shirt
una camiseta	T-shirt
una chaqueta	jacket
un cinturón	belt
una corbata	tie
una falda	skirt
un impermeable	raincoat
un jersey	pullover / jumper
unas medias	tights
unos pantalones	trousers
unos pantalones cortos	shorts
la ropa interior	underwear
unas sandalias	sandals
un traje	suit
un traje de baño	bathing suit
unos vaqueros	jeans
un vestido	dress
unas zapatillas deportivas	trainers
unos zapatos	shoes
la marca	make
el probador	changing room
la talla	size (clothes)
de moda	fashionable

Shopping for Clothes

¿Tiene esta falda en rojo, por favor?
Do you have this skirt in red, please?

Busco un vestido para una fiesta.
I'm looking for a dress for a party.

¿Me puede ayudar, por favor? ¿Dónde está el probador?
Can you help me, please? Where is the changing room?

¿Qué talla quiere usted?
What size would you like?

Perdone, ¿Puedo probarme este jersey?
Excuse me, can I try this jumper on?

Me quedo con esta pulsera de plata. ¡Es preciosa!
I'll take this silver bracelet. It's beautiful!

Accessories and Materials

un anillo	ring
un bolso	handbag
una bufanda	scarf
un casco	helmet
un cinturón	belt
un collar	necklace
unas gafas de sol	sunglasses
una gorra	cap
unos guantes	gloves
las joyas	jewellery
una pulsera	bracelet
un reloj	watch
un sombrero	hat

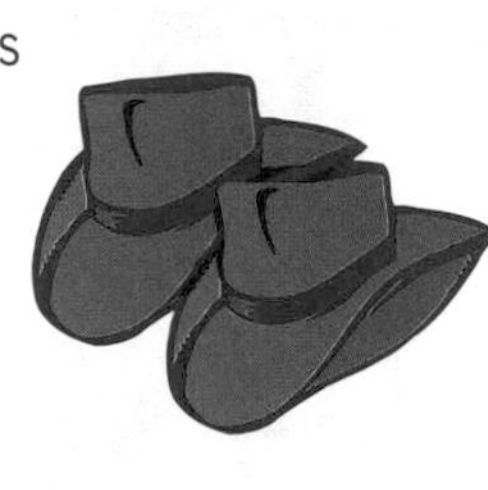

de algodón	(made of) cotton
de cuero / piel	(made of) leather
de lana	(made of) wool
de nylón	(made of) nylon
de oro	(made or) gold
de plata	(made of) silver
de seda	(made of) silk
de terciopelo	(made of) velvet

Colours

amarillo / a	yellow	**negro / a**	black
azul	blue	**rojo / a**	red
blanco / a	white	**rosa**	pink
gris	grey	**verde**	green
marrón	brown	**claro / a**	light
naranja	orange	**oscuro / a**	dark

✓ Maximise Your Marks

When describing clothes, make sure that all of your adjectives agree! For example: **las sandalias blancas** (white sandals); **la falda corta** (the short skirt); **los zapatos negros** (black shoes).

However, when a colour is made up of two words, it does *not* agree – it takes the masculine singular form: **la camisa rojo oscuro** (the dark red shirt); **las zapatillas deportivas azul claro** (light blue trainers).

Build Your Skills: Direct and Indirect Object Pronouns

Pronouns are used in place of nouns, to avoid having to repeat them. The word 'them' in the previous sentence is a direct object pronoun. Others include 'it' or 'me'. Indirect object pronouns in English are usually preceded by a word such as 'to' or 'from' – as in, for example, 'to me' or 'for him'.

- **Me gusta la falda. ¿La tiene en azul?**
  I like the skirt. Do you have it in blue?
  ('it' = direct object pronoun)
- **¿Me puede dar la falda roja?**
  Can you give (to) me the red skirt?
  ('me' = indirect object pronoun)

The direct object pronouns are:		The indirect object pronouns are:	
me	me	**me**	to/for me
te	you	**te**	to/for you
le/lo	him/it	**le**	to/for him / it
la	her/it	**le**	to/for her / it
le/la	you (**Vd.**)	**le**	to/for you (**Vd.**)
nos	us	**nos**	to/for us
os	you	**os**	to/for you
les/las	them (people)	**les**	to/for them
los	them (objects)	**les**	to/for them
les/las	you (**Vds.**)	**les**	to/for you (**Vds.**)

Note that in Spanish the pronoun must agree with the noun that it replaces. For example:

- **¿Tiene estas botas en negro? Sí, las tengo en negro y en marrón.**
  Do you have these boots in black? Yes, I have them in black and in brown.
- **¿Te gusta la camiseta? La compré para mi madre.**
  Do you like the T-shirt? I bought it for my mum.
- **¿Tienes el libro? No, no lo tengo.**
  Have you got the book? / No, I haven't got it.

If an indirect and a direct object pronoun are used together, in Spanish the indirect one must come first:

- **Mi madre me la dio.**
  My mum gave me it / gave it to me.

Object pronouns usually go before the verb, as above. However, with the immediate future or the present continuous, they either come before the verb or are attached to the end of the verbal phrase. For example:

- **Los voy a comprar. / Voy a comprarlos.**
  I'm going to buy them.
- **Los estoy comprando. / Estoy comprándolos.**
  I am buying them.

This also works for verbs such as **quiero** + infinitive:

- **Los quiero comprar. / Quiero comprarlos.**
  I want to buy them.

Pronouns are also attached to the end of an imperative:

- **¡Cómpralos!** Buy them!

Talking About Fashion

Me gusta estar a la moda y voy de compras cada fin de semana.
I like being fashionable and I go shopping every weekend.

Mis amigos prefieren comprar ropa de marca y siempre van de compras a las tiendas de diseño.
My friends prefer buying designer clothes and they always go to designer shops.

Mi madre compra la ropa en el supermercado porque dice que es más económica.
My mother buys clothes at the supermarket because she says that it's more economical.

A mi hermana le gusta la ropa alternativa y muchas veces la compra por Internet.
My sister likes alternative clothes and often buys them on the Internet.

? Test Yourself

What do these mean in English?

1. **¿Tiene la falda en la talla 38?**
2. **Ayer compré un collar de oro.**

How do you say these in Spanish?

3. a gold watch a woollen scarf black tights
4. Yesterday she bought a green dress and brown boots.

★ Stretch Yourself

Say or write these in Spanish:

1. Do you like my new shoes? I bought them because they were cheap!
2. Do you like this leather belt? I bought it for my brother.

Shops and Services

Shops

la agencia de viajes	travel agency
la biblioteca	library
la carnicería	butcher's
correos	post office
el estanco	tobacconist's
la farmacia	chemist's
la frutería	fruit shop
la hamburguesería	burger bar
la joyería	jeweller's
la librería	bookshop
el mercado	market
la panadería	baker's
la pastelería	cake shop
la peluquería	hairdresser's
la perfumería	perfume shop
la pescadería	fishmonger's
el quiosco	kiosk
el supermercado	supermarket
la tienda de comestibles	grocer's
la tienda de recuerdos	souvenir shop
la verdulería	greengrocer's
la zapatería	shoe shop
el centro comercial	shopping centre
los grandes almacenes	department stores

Boost Your Memory

As well as learning the names for different shops, you should also know the names of a few key people that work in them, as they may come up on reading and listening exams. Often, with shops that end in **–ería**, the word for the person working there will be the same but will end in **–ero / a** rather than **–ería**. In your revision notes, list the words in pairs for each shop and person working there. For example:

- **la peluquería** (hairdresser's)
  ➡ **el / la peluquero / a** (hairdresser)
- **la carnicería** (butcher's)
  ➡ **el / la carnicero / a** (butcher)

For most others, you can simply use the word for shop assistant: **el / la dependiente / a**.

Do not forget that the customer is **el / la cliente**.

Useful Vocabulary and Verbs

un billete	note (money)
un buzón	post box
la caja	checkout
un cajero automático	cash point
un carné de identidad	identity card
una carta	letter
un cheque (de viaje)	(traveller's) cheque
el dinero	money
un formulario	a form
una libra esterlina	pound sterling
una moneda	coin
una oferta especial	special offer
un paquete	parcel
un pasaporte	passport
las rebajas	the sales
un recibo	receipt
un sello	stamp
una tarjeta de crédito	credit card
una (tarjeta) postal	postcard
barato / a	cheap
caro / a	expensive
abrir	to open
buscar	to look for
cambiar	to change
cerrar	to close
comprar	to buy
costar	to cost
devolver	to return / give back / refund
encontrar	to find
envolver	to wrap up
enviar	to send
escoger	to choose
firmar	to sign
hacer cola	to queue
mandar	to send
pagar	to pay
pesar	to weigh
quejarse	to complain
rellenar	to fill in
sacar	to withdraw
vender	to sell

Build Your Skills: Demonstrative Adjectives and Pronouns

Demonstrative adjectives are words such as 'this', 'that', 'these' or 'those'. In Spanish they must agree with the noun that they go with.

Masc. Sing.	Fem. Sing.	Masc. Plural	Fem. Plural	English
este	esta	estos	estas	this / these
ese	esa	esos	esas	that / those
aquel	aquella	aquellos	aquellas	that / those… over there

Note that there are two words for 'that' – **ese** and **aquel**. The difference is that **aque**l refers to something further away:

- **ese supermercado** (that supermarket)
- **aquel supermercado** (that supermarket over there)
- **esta falda** (this skirt)
- **esa falda** (that skirt)
- **aquella falda** (that skirt over there)

By adding an accent to the demonstrative adjectives, they become demonstrative pronouns and can stand alone to mean 'this one' or 'that one':

¿Cuál banco? – ¿éste, ése o aquél?
Which bank, this one, that one or that one over there?

Esto, **eso** and **aquello** refer to an idea or anything that is not specifically mentioned by name:

¿Quién ha hecho esto?
Who did this?

Y ¿esto, qué es?
And what is this?

To refer to an unspecified thing ('something' / 'anything'), use the word **algo**; and to refer to an unspecified person ('someone' / 'anybody' etc.), use the word **alguien**:

Quiero comprar algo.
I want to buy something.

¿Alguien me puede ayudar?
Can anybody help me?

At the Shops

¿En qué puedo ayudarle? ¿Qué desea?
How can I help you? What would you like?

Está en la segunda planta / en la planta baja / en el sótano.
It's on the second floor / the ground floor / in the basement.

¿A qué hora se abre / se cierra?
What time does it open / close?

Quisiera mandar una carta a Inglaterra, por favor. ¿Cuánto es / cuesta?
I'd like to send a letter to England, please. How much is it / does it cost?

¿Puede decirme dónde hay un cajero automático?
Can you tell me where there is a cash point?

¿Dónde está la escalera? El ascensor no funciona.
Where are the stairs? The lift isn't working.

¿Tiene usted el recibo?
Have you got the receipt?

En cuanto al reloj, no puedo devolverle el dinero pero puedo cambiarlo.
Regarding the watch, I can't give you your money back but I can change it.

Querría / Quisiera un reembolso, por favor.
I'd like a refund, please.

Test Yourself

What do these mean in English?

1. **¿Dónde está la caja, por favor?**
2. **Ayer mandamos un paquete a mi abuela.**

How do you say these in Spanish?

3. My dad is a butcher, my mum is a baker.
4. I'm going to fill in a form at the post office.

Stretch Yourself

Say or write these in Spanish:

1. I like this postcard but not that one.
2. Which cake shop? This one or that one over there?
3. Do you want to go to this souvenir shop or that one?

Food and Drink

At the Market

la fruta	fruit
el albaricoque	apricot
la cereza	cherry
la ciruela	plum
la frambuesa	raspberry
la fresa	strawberry
el limón	lemon
la manzana	apple
el melocotón	peach
el melón	melon
la naranja	orange
la pera	pear
la piña	pineapple
el plátano	banana
el tomate	tomato
la uva	grape

la verdura	vegetable
el ajo	garlic
la cebolla	onion
el champiñón	mushroom
la col	cabbage
las coles de bruselas	sprouts
la coliflor	cauliflower
el espárrago	asparagus
las espinacas	spinach
los guisantes	peas
las judías verdes	green beans
la lechuga	lettuce
las legumbres	vegetables
la patata	potato
el pepino	cucumber
el pimiento	pepper
la zanahoria	carrot

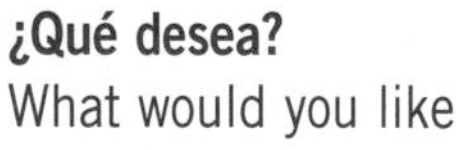

¿Qué desea?
What would you like?

¿Cuánto cuestan las manzanas?
How much are the apples?

Un euro cincuenta el kilo.
One euro fifty a kilo.

Pues, deme dos kilos por favor y medio kilo de uvas.
Well, give me two kilos please and half a kilo of grapes.

¿Algo más? Anything else?

No, nada más, gracias. ¿Cuánto es?
No, nothing else thanks. How much is it?

At the Supermarket

una lata de atún	a tin of tuna
un cartón de leche	a carton of milk
una botella de vino tinto	a bottle of red wine
un kilo de tomates	1 kilo of tomatoes
doscientos gramos de queso	200 grammes of cheese
cinco lonjas de jamón de york	5 slices of cooked ham
un bote de café	a jar of coffee
dos barras de pan	2 loaves of bread
un paquete de azúcar	a packet of sugar
una docena de huevos	a dozen eggs
un litro de agua	1 litre of water
una caja de galletas	a box of biscuits
un paquete de mantequilla	a pack of butter

¿En qué puedo servirle?
How can I help you?

Póngame quinientos gramos de queso y dos kilos de jamón serrano, por favor.
I'll have (lit. 'put me') 500 grammes of cheese and 2 kilos of Serrano ham, please.

Aquí tiene. ¿Algo más?
Here you are. Anything else?

Sí, deme seis chuletas de cerdo también.
Yes, give me six pork chops as well.

Lo siento, sólo quedan cuatro. ¿Las quiere?
I'm sorry, there are only four left. Do you want them?

Sí, me las llevo, gracias.
Yes, I'll take them, thank you.

Boost Your Memory

The list of food vocabulary is endless and may seem very daunting to remember! A good idea is to group the words together into smaller lists, i.e., fruit, vegetables, drinks, meat, grocery items, desserts, etc., and then learn them one group at a time. Start with any cognates (words that look similar to the English and mean the same thing) and learn those first. Cognates such as **el tomate, el limón, la coliflor** will be the easiest to remember. Then move on to the more difficult words such as **la zanahoria**, etc. Make sure that you can spell the cognates correctly, as well as pronounce them properly in Spanish.

Drinks

las bebidas	drinks
el agua	water
el agua mineral (con / sin gas)	mineral water (fizzy / still)
el café (con leche / solo)	coffee (with milk / black)
la cerveza	beer
la coca-cola	cola
la gaseosa	lemonade / pop
la leche	milk
la limonada	lemonade
la naranjada	orangeade
el refresco	soft drink
el té	tea
el tinto	red wine
el vino (blanco)	(white) wine
el zumo de fruta	fruit juice
el zumo de manzana	apple juice

Tengo sed. ¿Quieres tomar algo?
I'm thirsty. Do you want something to drink?

Sí, para mí, un café con leche y agua mineral sin gas, por favor.
Yes, I'll have (for me) a white coffee and still mineral water, please.

Adverbs

Adverbs describe how an action or something is done, for example 'quickly', 'badly', 'always'. They often end in '–ly' in English. In Spanish, they are formed by adding **–mente** to the feminine form of the adjective. For example:

- **lento / a** (slow) ➡ **lentamente** (slowly)
- **probable** (probable) ➡ **probablemente** (probably)

Of course, there are irregular words that do not follow this pattern:

ahora	now
allí	there
a menudo	frequently / often
aquí	here
algunas veces	sometimes
bastante	enough
bien	well
demasiado	too / too much
deprisa	fast / quick
mal	badly
mucho	a lot
muchas veces	often
siempre	always
ya	already

¡Mi padre siempre bebe lentamente!
My dad always drinks slowly!

Build Your Skills: Comparative and Superlative of Adverbs

Comparatives of adverbs are formed in the same way as comparatives of adjectives, using **más...que**, **menos...que**, etc.:

- **Mi padre bebe más lentamente que mi madre.**
  My dad drinks more slowly than my mum.
- **Compro fruta en el mercado más a menudo que en el supermercado.** *or* **Compro fruta más a menudo en el mercado que en el supermercado.**
  I buy fruit in the market more often than in the supermarket.

Take care with the superlative of adverbs: the definite article (**el**, **la**, etc.) cannot be used and therefore the superlative often looks like a comparative:

- **De toda la fruta, la que compro más a menudo es la piña.**
  Of all fruit, the one I buy (the) most often is pineapple.

Using adverbs, especially in comparing different people and things, is a good way to extend your sentences and so gain more marks in the exam.

? Test Yourself

What do these mean in English?

1. **Póngame ciento cincuenta gramos de jamón.**
2. **Para mí un zumo de naranja, por favor.**

How do you say these in Spanish?

3. Give me a kilo of bananas and half a kilo of green beans, please.
4. My friend always eats too much.

★ Stretch Yourself

Say or write these in Spanish:

1. The waiter speaks more quickly than I (do).
2. In the bar I order (ask for) a white coffee more frequently than a soft drink.

The Perfect and Pluperfect Tenses

Build Your Skills: The Perfect Tense

The perfect tense is used to say what someone *has done* or what *has happened*. For example: 'I have eaten', 'She has walked', 'We have been', etc. By using the perfect tense in your writing and speaking, you can give more detail about the past and therefore gain more marks.

In Spanish, the perfect tense is formed by using the present tense of the verb **haber** and a past participle. In English, past participles of regular verbs usually end in '–ed': 'walked', 'listened', 'studied', etc. In Spanish, the endings **–ado** (**–ar** verbs) or **–ido** (**–er** and **–ir** verbs) are added to the stem of the verb.

Haber	(To have) as an auxiliary verb	+ past participle
He	I have	
Has	You have (**tú**)	
Ha	He / She / It has; You have (**Vd.**)	**hablado** (spoken) + **comido** (eaten) **salido** (gone out)
Hemos	We have	
Habéis	You have (**vosotros**)	
Han	They have; You have (**Vds.**)	

He escuchado **las noticias en la radio.**
I have listened to the news on the radio.

¡El perro ha comido **la cena!**
The dog has eaten the dinner!

As usual, there are a number of irregular past participles, which have to be learned separately. The common irregular ones are:

abrir (to open) → **abierto**
He abierto **mi libro en la página 10.**
I have opened my book at page 10.

cubrir (to cover) → **cubierto**
Las nubes han cubierto **la montaña.**
The clouds have covered the mountain.

decir (to say) → **dicho**
¡Yo no he dicho **nada!**
I haven't said anything!

descubrir (to discover) → **descubierto**
¡Hemos descubierto **una tienda de ropa fantástica!**
We have discovered a fantastic clothes shop!

escribir (to write) → **escrito**
He escrito **una canción en español.**
I have written a song in Spanish.

hacer (to do / make) → **hecho**
He hecho **el esquí en Francia.**
I have done (been) skiing in France.

morir (to die) → **muerto**
Desafortunadamente, ha muerto.
Unfortunately, she has died.

poner (to put) → **puesto**
¿Dónde he puesto **el periódico?**
Where have I put the newspaper?

romper (to break) → **roto**
Mi hermano ha roto **el nuevo videojuego.**
My brother has broken the new videogame.

ver (to see) → **visto**
¿Has visto **mi flauta?**
Have you seen my flute?

volver (to return) → **vuelto**
No he vuelto **al gimnasio.**
I haven't gone back / returned to the gym.

Nothing comes between haber and the past participle. All pronouns and negatives come before haber:

No he hablado **con mi madre hoy.**
Me he levantado **muy temprano para hacer footing.**
I haven't spoken to my mother today.
I have got up very early to go jogging.

Build Your Skills: The Pluperfect Tense

The pluperfect tense describes what *had happened* or what someone *had done*. It is one tense further back in time than the preterite and the perfect tense. So if you are describing two things that happened in the past, and one happened before the other, you use the pluperfect for the event that happened first. If you are narrating something in the past, you use the pluperfect to imply that something happened earlier (i.e. it had already happened).

Like the perfect, the pluperfect is formed by using the verb **haber** and a past participle, but this time **haber** is in the imperfect tense:

Haber	To have (as an auxiliary)	+	past participle
Había	I had		
Habías	You had (**tú**)		**hablado** (spoken)
Había	He / She / It had; You had (**Vd.**)	+	**comido** (eaten)
Habíamos	We had		**salido** (gone out)
Habíais	You had (**vosotros**)		
Habían	They had; You had (**Vds.**)		

Había visto el programa en la televisión
He had watched the programme on the television

Había hablado con mi hermano antes de llegar.
I had spoken to my brother before arriving / before I arrived.

Habían salido con sus amigos la semana anterior.
They had gone out with their friends the week before.

Cuando Miguel llegó a casa, María había comido ya.
When Miguel arrived home, Maria had already eaten.

Build Your Skills: Saying 'I Have / Had Just…'

To say 'I have just' + past participle, for example 'I have just finished', in Spanish you do not use the perfect tense, you use the verb **acabar** + **de** + infinitive:

- **Acabo de terminar.**
  I have just finished.
- **¡Acabo de ver una película buenísima!**
  I have just watched a brilliant film!

To say 'I had just…', you use the verb **acabar** in the imperfect + **de** + infinitive:

- **Acababa de terminarlo cuando me llamó mi madre.**
  I had just finished it when my mum called me.

✓ Maximise Your Marks

The perfect and pluperfect tenses are higher level so, if possible, try and include an example of one in your speaking or writing, or even both in the same sentence, to really impress with a complex sentence!

- **Anteriormente, había pensado que era demasiado difícil para mí, pero recientemente he cambiado mi manera de pensar.**
  I had previously thought that it was too difficult for me but I have changed my mind recently.

? Test Yourself

What do these mean in English?

1. **Esta mañana he desayunado cereales con leche.**
2. **Me había preguntado antes si quería ir al cine.**

How do you say these in Spanish?

3. They have written a letter to my teacher.
4. I had already visited Seville.

★ Stretch Yourself

Say or write these in Spanish:

1. In the end, the passport that he had lost was in his bag!
2. She had just eaten her lunch when her father arrived home.

Eating Out

At the Restaurant

La carta / El menú	Menu
Primer plato	*First course*
el chorizo	garlic sausage
la ensalada	salad
los espaguetis	spaghetti
el fiambre	sliced meat (e.g. salami)
el gazpacho	cold vegetable soup
la salchicha	sausage
la sopa de verduras	vegetable soup
la tortilla española	Spanish omelette
Segundo plato	*Second course*
la carne	meat
las albóndigas	meatballs
el bistec	steak
la carne de cerdo	pork
la carne de vaca	beef
el cocido	stew
el cordero	lamb
el pollo	chicken
la ternera	veal
el tocino	bacon
el pescado	fish
el bacalao	cod
la merluza	hake
las sardinas	sardines
la trucha	trout
los mariscos	seafood
los calamares	squid
las gambas	prawns
la paella de mariscos	seafood paella
Postres	*Desserts*
el arroz con leche	rice pudding
el flan	crème caramel
el helado de vainilla	vanilla ice-cream
la tarta de chocolate	chocolate cake

Build Your Skills: The Definite Article with 'Gustar' and 'Encantar'

When discussing likes and dislikes using **gustar** or **encantar**, remember to use the definite article (**el / la / los / las**) with the noun:

Me gustan mucho las **gambas.**
I really like prawns.

No me gusta el **bistec.**
I don't like steak.

¡Me encanta el **chocolate!**
I love chocolate!

However, when using other verbs to talk about food, it is not necessary to use the definite article:

- **Como fruta y bebo agua todos los días.**
  I eat fruit and I drink water every day.
- **De segundo, suelo comer pescado.**
  For the second course, I usually have fish.

✓ Maximise Your Marks

Listening and reading exams will often contain extra bits of information that you do not need in order to answer the question. These are called 'distractors'. You have to try to work out exactly which is the correct information for the answer. For example, two people could be discussing what to order in a restaurant, but could also mention some things that they like in general. That does not necessarily mean that they are going to order those things!

As well as listening to the vocabulary, listen out for key verbs and phrases such as **quiero** or **quisiera** (I would like), **para mí** (for me...), etc., so that you know that you have got the correct information. Then listen to the end of the passage to make sure!

Useful Verbs

almorzar	to have lunch
apetecer	to feel like / fancy
beber	to drink
cenar	to have dinner (evening meal)
comer	to eat
desayunar	to have breakfast
emborracharse	to get drunk
hacer / preparar	to make / prepare
merendar	to have a snack
pedir	to order / ask for
probar	to try
saber	to taste
seguir una dieta / estar de régimen	to be on a diet
tener hambre	to be hungry
tener sed	to be thirsty
tomar	to have (food / drink)

¿Qué va a tomar?
What are you going to have?

De primero, quisiera el gazpacho y, de segundo, la paella de mariscos.
For the first course, I'd like the gazpacho and, for the second course, the seafood paella.

Y ¿para beber?
And to drink?

Quiero vino blanco, por favor.
I'd like white wine, please.

La cuenta, por favor.
The bill, please.

Breakfast

el desayuno	breakfast
los cereales	cereal
los churros	fritters
las galletas	biscuits
el huevo	egg
la mantequilla	butter
la mermelada	jam
la mermelada de naranja	marmalade
la miel	honey
el pan	bread
el pan tostado / la tostada	toast
el panecillo	breadroll
el yogur	yogurt

¿Qué sueles desayunar?
What do you usually have for breakfast?

Pues, normalmente tomo un vaso de leche con tostadas.
Well, I usually have a glass of milk with toast.

Yo desayuno cereales durante la semana pero el sábado me gusta tomar churros con chocolate caliente.
I have cereal for breakfast during the week, but on a Saturday I like to have fritters and hot chocolate.

? Test Yourself

What do these mean in English?

1. **Para mí de primero, la tortilla española, por favor.**
2. **Suelo desayunar pan tostado con mermelada de fresa.**

How do you say these in Spanish?

3. I'll have garlic sausage for the first course and stew for the second.
4. Do you fancy seafood or meat?

★ Stretch Yourself

Say or write these in Spanish:

1. I love seafood paella but I don't like meatballs at all!
2. I really like steak but I don't like chicken at all!

Healthy Living

Health and Fitness Vocabulary

el alimento	food
el apetito	appetite
el almuerzo	lunch
el cansancio	tiredness
la cena	dinner / evening meal
la comida	food / lunch
la comida basura	junk food
la comida rápida	fast food
delgado / a	slim
la dieta sana / malsana	healthy / unhealthy diet
la dieta equilibrada	balanced diet
la depresión	depression
el ejercicio físico	physical exercise
el estrés	stress
las golosinas	sweet things
gordo / a	fat (adjective)
la grasa	fat
los granos	spots
perjudicial	harmful
la salud	health
saludable	healthy
sano / a	healthy

Alcohol

adictivo / a	addictive
el alcohol	alcohol
el alcoholismo	alcoholism
el alcohólico	alcoholic
borracho / a	drunk
con moderación	in moderation
la tentación	temptation

En mi opinión, el alcohol puede ser adictivo para alguna gente, y eso es muy grave.
In my opinion, alcohol can be addictive for some people, and that is very serious.

Cuando sea mayor, no voy a beber alcohol porque puede ser perjudicial para la salud.
When I am older, I am not going to drink alcohol because it can be harmful to your health.

Mis amigos y yo nos emborrachamos los fines de semana y no veo nada malo en eso.
My friends and I get drunk at weekends and I don't see anything wrong with that.

Useful Verbs

correr el riesgo	to run the risk
drogarse	to take drugs
emborracharse	to get drunk
entrenarse	to train
estar en forma	to be fit / healthy
evitar	to avoid
fumar	to smoke
hacer ejercicio	to exercise
inyectarse	to inject
mantenerse en forma	to keep fit
estar en forma	to be fit
oler	to smell
perder peso	to lose weight
relajarse	to relax
respirar	to breathe

Creo que estoy en forma. Intento comer una dieta equilibrada, o sea carne, pescado, fruta, verduras, etc., y practico el deporte tres veces a la semana.
I think I am fit. I try to eat a balanced diet, that's to say meat, fish, fruit, vegetables, etc., and I do sport three times a week.

Para llevar una vida más sana, no deberías comer demasiada comida rápida.
To lead a healthier life, you should not eat too much fast food.

Cuando era más joven, comía muchas golosinas y comida que contiene grasa.
When I was younger, I used to eat lots of sweet things and food that contains fat.

Hoy en día estoy intentando seguir una dieta más saludable.
These days I am trying to follow a healthier diet.

En el futuro, no voy a beber alcohol porque puede ser perjudicial para la salud.
In the future, I am not going to drink alcohol because it can be harmful to your health.

Build Your Skills: Expressing Strong Opinions

There are straightforward phrases you can use to give opinions, such as **en mi opinión**, **creo que**, **pienso que**, etc. However, when you are talking about topics such as drugs and smoking, you could express your opinions in a slightly stronger way with phrases such as these:

- **Estoy a favor de** I am in favour of
- **Estoy en contra de** I am against
- **No estoy ni a favor ni en contra de** I'm neither for nor against
- **Desde mi punto de vista** From my point of view
- **A mi modo de ver** In my view
- **Está claro que** It's clear that
- **(No) estoy de acuerdo con** I (dis)agree with

No estoy ni a favor ni en contra de fumar y, desde mi punto de vista, todo el mundo tiene que tomar sus propias decisiones.
I'm neither for nor against smoking and, from my point of view, everyone has to make their own decisions.

Using a variety of structures in your speaking and writing will enrich your language and earn you more marks in the exam.

✓ Maximise Your Marks

In listening and reading exams, use all possible clues to help you decide on your answer. Look out for opinion phrases to see whether someone is being positive or negative about something; tone of voice; time expressions in the past, present and future, etc. Even if you do not understand all of the vocabulary, use what you do understand to make an educated guess.

Drugs

la adicción	addiction
el / la adicto / a	addict
la cocaína	cocaine
la droga blanda / dura	soft / hard drug
el / la drogadicto / a	drug addict
el hábito	habit
la inyección	injection
la muerte	death
peligroso	dangerous
la rehabilitación	rehabilitation
el SIDA	Aids

Mis amigos toman drogas blandas para relajarse.
My friends take soft drugs to relax.

No quiero probar las drogas porque me dan miedo.
I don't want to try drugs because they scare me.

Alguna gente toma drogas a causa de la presión por el grupo paritario.
Some people take drugs because of peer pressure.

Smoking

asqueroso / a	disgusting
el cigarrillo	cigarette
el fumador	smoker
el fumar pasivo	passive smoking
el humo	smoke
los pulmones	lungs
el tabaco	tobacco
el tabaquismo	smoking habit

Para mí, el humo de los cigarrillos es asqueroso.
For me, cigarette smoke is disgusting.

Fumar es muy perjudicial para los pulmones.
Smoking is very harmful for the lungs.

Mi padre fuma veinte cigarillos al día.
My dad smokes 20 cigarettes a day.

? Test Yourself

What do these mean in English?

1. **Estoy totalmente en contra de las drogas. ¡Son muy peligrosas!**
2. **Fumar es muy perjudicial para la salud.**

How do you say these in Spanish?

3. To lead to healthier life, you should not smoke.
4. You have to eat well and exercise to be fit.

★ Stretch Yourself

Say or write these in Spanish:

1. It's clear that taking drugs is dangerous.
2. In my view, drugs are a very serious problem these days.

Illness and Accidents

The Body and Saying What Hurts

la boca	mouth	**el estómago**	stomach
el brazo	arm	**la garganta**	throat
la cabeza	head	**la mano**	hand
la cara	face	**las muelas**	teeth
el corazón	heart	**la nariz**	nose
el cuello	neck	**el oído**	inner ear
el cuerpo	body	**el ojo**	eye
el dedo	finger	**la oreja**	ear
el diente	tooth	**el pie**	foot
la espalda	back	**la pierna**	leg

To say that something hurts, in Spanish, you use the verb **doler**. This is a radical-changing verb (**o** > **ue**) and it works the same way as **gustar**:

- **Me duele la cabeza.**
  My head hurts. (lit. My head is hurting to me.)
- **¿Te duele la pierna?**
  Does your leg hurt? / Is your leg hurting (to you)?

For plural nouns, use **duelen**:

- **Me duelen las muelas.**
  My teeth hurt.
- **Le duelen los brazos.**
  His / Her arms hurt.

Another way to say that something hurts, or that you have a pain somewhere, is to use **tener dolor de…**:

- **Tengo dolor de garganta.**
  I have a sore throat (lit. a pain of the throat).
- **Tiene dolor de estómago.**
  He / She has stomach ache.

Other Expressions Using 'Tener'

As well as **tener dolor**, there are many other useful expressions in Spanish that use the verb **tener** where in English we tend to use the verb 'to be'. For that reason, they will enrich your language if you use them in the exam!

Tengo...	
dieciséis años	I am 16 (years old)
calor	I am hot
éxito	I am successful
frío	I am cold
hambre	I am hungry
miedo	I am frightened
prisa	I am in a hurry
razón	I am right
sed	I am thirsty
suerte	I am lucky

Tengo que ir. I have to go.

Discussing Accidents

¿Qué pasó ayer en la plaza?
What happened yesterday in the square?

Hubo un accidente, una colisión entre dos coches.
There was an accident, a collision between two cars.

¡Ay, qué horror! ¿Hubo heridos?
Oh, how awful! Were there any injuries?

Sí, llevaron a uno de los conductores al hospital porque se había roto la pierna.
Yes, they took one of the drivers to hospital because he had broken his leg.

The Present Continuous

The present continuous tense describes what *is happening* at this moment. In Spanish, it is formed by using the appropriate part of the verb **estar** in the present tense and a gerund. In English, the gerund takes the form '–ing': 'speaking'; 'walking'; 'running', for example. In Spanish, the gerund is formed by adding **–ando** (**–ar** verbs) or **–iendo** (**–er** and **–ir** verbs) to the stem of the verb: **hablando** (speaking), **comiendo** (eating), etc.

Here are some examples of the present continuous tense:

- **Estoy hablando.**
  I am speaking.
- **Estamos viendo una película.**
  We are watching a film.
- **Hoy voy a ir al medico porque estoy tosiendo mucho.**
  I am going to go to the doctor's today because I'm coughing a lot.

Saying How You Feel

cortar(se)	to cut (oneself)
desmayarse	to faint
doler	to hurt
estar bien	to feel well
estar constipado	to have a cold
estar mal	to feel ill
estar resfriado	to have a cold
guardar cama	to stay in bed
herirse	to injure oneself
marearse	to get dizzy / seasick
picar	to bite / sting
quemar(se)	to burn (oneself)
sentirse	to feel
toser	to cough
vomitar	to vomit

¿Qué tal te sientes?
How do you feel?

Pues, tengo dolor de espalda y me duele mucho la cabeza.
Well, I have backache and my head hurts a lot.

Hoy no me siento bien. Tengo tos y fiebre también.
I don't feel well today. I have a cough and a temperature as well.

Ayer mi hermana se cortó la pierna y tuvo que ir al médico.
Yesterday, my sister cut her leg and had to go to the doctor's.

Build Your Skills: The Imperfect Continuous

The imperfect continuous tense describes what *was happening* at a particular moment in the past. It is formed with the appropriate part of **estar** and a gerund, only this time **estar** is in the imperfect tense.

Here are some examples of the imperfect continuous tense:

- **¿Qué estabas haciendo?**
  What were you doing?
- **Estaba escribiendo un ensayo.**
  I was writing an essay.
- **Mi hermana estaba vomitando todo el día.**
  My sister was vomiting all day long.

Boost Your Memory

Having lots of different tenses to learn can be confusing. Design yourself a grid or a spreadsheet and fill it in with a few examples of regular verbs in different tenses and the main irregular verbs. Try colour-coding it in a way that will help you to see patterns more easily. Or try recording yourself saying the verbs and listen back as often as possible. Find a strategy that works for you!

Test Yourself

What do these mean in English?

1. **Me siento mal – me duele mucho la garganta.**
2. **Ayer me quemé el dedo y me dolió bastante.**

How do you say these in Spanish?

3. My legs and my feet hurt!
4. I have to stay in bed because I have a temperature.

Stretch Yourself

Say or write these in Spanish:

1. I am staying in bed because I don't feel well.
2. Yesterday, I was cooking in the kitchen when I fainted.

Celebrations and Special Events

Special Events

el Año Nuevo Chino	Chinese New Year
el cumpleaños	birthday
el día de Año Nuevo	New Year's Day
el día de Navidad	Christmas Day
el día de Reyes	Epiphany (6 January)
el día del santo	Saint's Day
el Diwali	Diwali
el festival de Eid	Eid
la fiesta de Januka	Hanukkah
las Navidades	Christmas
la Nochebuena	Christmas Eve
la Pascua	Easter
la Nochevieja	New Year's Eve
la Semana Santa	Easter / Holy Week

Celebration Verbs and Sentences

adornar	to decorate
cantar	to sing
celebrar	to celebrate
dar	to give
esperar con ganas / ilusión	to look forward to
felicitar	to congratulate
mandar	to send
ofrecer	to offer / give
recibir	to receive
regalar	to give as a present
rezar	to pray
tener lugar	to take place

¿Cómo celebras las Navidades en tu casa?
How do you celebrate Christmas in your house?

Normalmente adornamos un árbol de Navidad y también montamos un belén. La Nochebuena cenamos pescado y luego vamos todos a la misa del gallo.
Usually, we decorate a Christmas tree and set up a Nativity scene. On Christmas Eve we eat fish and then we all go to Midnight Mass.

Todo el mundo da regalos y la familia pasa mucho tiempo junta comiendo, jugando, a veces cantando villancicos y generalmente pasándolo bien.
Everyone gives presents and the family spends a lot of time together eating, playing, sometimes singing Christmas carols and generally having a good time.

En España los niños reciben los regalos el día de Reyes, que es el 6 de enero.
In Spain, children receive their presents on the Epiphany, which is the 6th of January.

Para celebrar la Nochevieja, se bebe champán y a las doce se come doce uvas, que es típico en España.
To celebrate New Year's Eve, people drink champagne and at 12 o'clock they eat 12 grapes, which is the tradition in Spain.

¡A mi me encanta la Semana Santa! El año pasado fui a Sevilla para ver las procesiones y ¡fue muy impresionante!
I love Holy Week! Last year I went to Seville to see the processions and it was really impressive!

Indefinite Adjectives

Indefinite adjectives are words such as **cada** (each / every), **alguno** (a / some) and **otro** (other / another). These words can cause problems when it comes to adjectival agreement:

The word **cada** never changes:

- **Cada Nochebuena vamos a casa de mis abuelos.**
  Every Christmas Eve we go to my grandparents' house.

You must not use the indefinite article when using the adjective otro, and it must agree with the noun:

- **Vamos a otra fiesta el sábado.**
  We are going to another party on Saturday.

The words **todo** (all / every) and **mismo** (same) also agree with the noun:

- **Todos los años celebramos mi cumpleaños en el mismo restaurante**.
  Every year we celebrate my birthday in the same restaurant.

Alguno and **ninguno** (not any / no) lose their final **–o** if they are placed in front of a masculine singular noun:

- **Algún día iré a Sudamérica para celebrar la Semana Santa.**
  Some day, I will go to South America to celebrate Holy Week.

✓ Maximise Your Marks

Accuracy in speaking and writing is very important. This is also one of the most common areas where marks are lost in the exam. Check your work as often as possible to try to spot any common mistakes such as adjectival agreements, misspelt cognates, omitted accents, etc.

Build Your Skills: Adverbial Phrases

Adverbial phrases help to enrich your writing and speaking and make your language more interesting – and so gain you more marks. Try using **muchas veces** (often), **dentro de poco** (soon), **por todas partes** (everywhere) or **en otra parte** (elsewhere). Also, phrases such as **con entusiasmo** (enthusiastically), **en secreto** (secretly) and **de manera sorprendente** (surprisingly) all help to 'liven up' your sentences!

En mi casa, esperamos con ilusión **la fiesta de Diwali, el festival de las luces. Es cuando celebramos el año nuevo hindú y me encanta.** Dentro de poco**, decoraremos la casa con lámparas de colores** por todas partes **y prepararemos mucha comida porque vendrán muchos amigos y familia.**
In my house, we are looking forward to Diwali, the festival of lights. It is when we celebrate the Hindu New Year and I love it. Soon we will decorate the house with coloured lights everywhere and we will cook a lot of food because many friends and family will come.

? Test Yourself

What do these mean in English?

1. **Me regalaron una chaqueta de cuero.**
2. **Mi madre preparó una cena muy rica.**

How do you say these in Spanish?

3. I celebrate my birthday with my friends.
4. I am looking forward to Chinese New year.

★ Stretch Yourself

Say or write these in Spanish:

1. We often celebrate Easter at my aunt and uncle's house.
2. I am going to celebrate my birthday soon.

Practice Questions

Complete these exam-style questions to test your skills and understanding. Check your answers on pages 93–94. You may wish to answer these questions on a separate piece of paper.

Reading

1 Read the following blogs and answer the questions.

Hola, en mi tiempo libre lo que me gusta hacer es practicar el deporte. Juego al baloncesto tres veces a la semana en el colegio y los sábados voy a la piscina con mis amigos. Mi hermana practica la gimnasia pero a mi modo de ver es aburrida. También soy miembro del equipo de atletismo y suelo entrenarme los lunes y los jueves. Ana
Yo soy futbolista y la semana pasada marqué dos goles para mi equipo. ¡Qué guay! Soy hincha del Real Madrid y voy a los partidos con mi padre. También, en mi tiempo libre suelo ir a casa de un amigo y jugamos con el ordenador o damos un paseo con su perro. No leo nunca porque prefiero ver la tele o escuchar música. Rafa
A mí me encanta ir de pesca con mi hermano y mi padre. Hay un lago cerca de mi pueblo y normalmente pasamos los domingos allí. Es muy tranquilo y llevamos algo para comer y beber. A veces voy al centro para ver una película con mis amigos pero no vamos todas las semanas porque depende de si tenemos suficiente dinero. Manolo

a) What does Ana do three times a week? .... (1)

b) Who thinks gymnastics is boring? .... (1)

c) What does Rafa do with his father? .... (1)

d) Who goes swimming with friends? .... (1)

e) What does Ana do on Mondays and Thursdays? .... (1)

f) What does Manolo take when he goes fishing? .... (1)

g) Where does he go fishing? .... (1)

h) Why doesn't Manolo go to the cinema every week? .... (1)

2 Read the opinions (**A** to **C**) of three people on an article about alcohol. Choose the correct summary in English for each of the opinions and write in the correct letter.

A **En mi clase, la mayoría de los chicos beben alcohol y nos divertimos mucho cuando salimos juntos. No veo ningún problema.** ☐ (1)

B **Empecé a beber un poco con mis amigos los fines de semana en la calle o en el parque.** ☐ (1)

C **Conozco a gente que bebe porque los amigos lo hacen. Realmente no quieren hacerlo pero no quieren parecer aburridos.** ☐ (1)

a) Alcohol is dangerous for your health

b) Peer pressure to drink

c) My first drinking experiences

d) Alcohol is fine in small quantities

e) Drinking alcohol is no big deal

Speaking

3 You are discussing shopping habits with your Spanish friend. Prepare to speak in Spanish about each of the points below.

a) Where you tend to go shopping and why

b) Who you prefer to go shopping with

c) Where you prefer to go to buy clothes

d) How you spent your day the last time you went shopping

e) What you think of department stores and why

f) What your ideal shopping experience would be like

(10)

Writing

4 You have been asked to write an article about your lifestyle for a Spanish school magazine. Write in Spanish about each of the following:

- Your eating and drinking habits
- What sport and exercise you do now
- How your diet and exercise has changed from when you were younger
- Whether you consider yourself to be fit and healthy and why
- What changes you need to make to your lifestyle in the future
- Your opinions about smoking, drugs and alcohol

(15)

How well did you do?

1–12 Try again	13–22 Getting there	23–30 Good work	31–36 Excellent!

School and School Subjects

School Subjects

las asignaturas	subjects
el alemán	German
la biología	biology
las ciencias	sciences
el comercio	business studies
el dibujo	art
la educación física	PE
el español	Spanish
la física	physics
el francés	French
la geografía	geography
la historia	history
los idiomas	languages
la informática	ICT
el inglés	English
el italiano	Italian
las matemáticas	maths
la música	music
la religión	RE
el teatro	drama
la tecnología	technology

Positive and Negative Descriptions

Use these adjectives to give positive opinions:

creativo / a	creative
divertido / a	fun / amusing
entretenido / a	entertaining / amusing
fácil	easy
interesante	interesting
práctico	practical
útil	useful

Use these adjectives to give negative opinions:

aburrido / a	boring
complicado / a	complicated
difícil	difficult
injusto	unfair
inútil	useless

Useful Verbs

aburrirse	to be bored
aprender	to learn
apoyar	to support
aprobar (un examen)	to pass (an exam)
castigar	to punish
comenzar	to begin
comprender	to understand
contestar	to answer
corregir	to correct
deber	to have to
deletrear	to spell
dibujar	to draw
empezar	to start
enseñar	to teach
entender	to understand
escuchar	to listen
escribir	to write
estudiar	to study
explicar	to explain
faltar	to be absent
odiar	to hate
olvidarse de	to forget
pasar lista	to call the register
pedir permiso	to ask for permission
preferir	to prefer
preguntar	to ask
repasar	to revise
sacar buenas / malas notas	to get good / bad marks
suspender	to fail

Boost Your Memory

There is a lot of vocabulary to remember when it comes to talking about school, so try different strategies to help you memorise as much as possible. For example, with school subjects, you could try giving yourself 3 minutes to write down as many different subjects as you can, with the definitive article (**el** / **la** / **los** / **las**), and then check them for accuracy. See which ones you could not remember and then repeat the exercise starting with the ones you had forgotten. Do not just read through the words once and expect to remember them all. You have to make an effort to memorise them!

Impersonal Verbs

Some verbs in Spanish, including some verbs of opinion, are used impersonally. An example is **me gusta el español** (I like Spanish). In the English, the subject is 'I' and the verb is personal, but in Spanish the subject is **el español** and the verb is impersonal (the phrase literally means 'Spanish pleases / is pleasing to me'). In Spanish the verb is usually in the third person singular or plural (it / they). The subject or person in the English construction becomes an indirect object pronoun (to me) and comes before the verb. So, instead of changing the ending of the verb to show the person, in Spanish you use the correct indirect object pronoun:

Me **gusta(n)**	I like
Te **gusta(n)**	You like (**tú**)
Le **gusta(n)**	He / She likes; You like (**Vd.**)
Nos **gusta(n)**	We like
Os **gusta(n)**	You like (**vosotros**)
Les **gusta(n)**	They like; You like (**Vds.**)

Me gusta el inglés.
I like English.

Te gustan las ciencias.
You like science.

Note that **gusta** becomes **gustan** with a plural noun. Remember also that the definite article (**el** / **la** / **los** / **las**) must be used.

Other verbs of opinion that follow this pattern are **encantar** (to love) and **interesar** (to interest):

Me encanta la historia.
I love history.

Le interesan los idiomas.
He is interested in languages.

Other impersonal verbs are:

doler	to have a sore.../ to ache
faltar	to lack / to be short of
hacer falta	to need
quedar	to have...left
sobrar	to exceed / be left over

Me encanta el español porque es muy interesante y divertido.
I love Spanish because it's very interesting and fun.

Se me da bien el inglés pero se me dan mal las ciencias.
I'm good at English but I'm not good at science.

A mi hermano le interesan mucho las matemáticas pero yo las odio porque son muy difíciles.
My brother is really interested in maths but I hate it because it's very difficult.

Talking About Your Subjects

Mi asignatura preferida es la historia porque el profesor es muy interesante.
My favourite subject is history because the teacher is very interesting.

No aprendo mucho en la clase de música porque la encuentro muy difícil.
I don't learn much in my music lesson because I find it very difficult.

El año pasado estudié informática pero la dejé porque era muy aburrida.
Last year I studied ICT but I dropped it because it was very boring.

Preferiría estudiar la educación física porque es divertida y práctica.
I would prefer to study PE because it's fun and practical.

? Test Yourself

What do these mean in English?

1. **No me gusta la física porque no entiendo nada.**
2. **Voy a suspender la música porque siempre saco malas notas.**

How do you say these in Spanish?

3. I want to get good marks in English.
4. I would love to study Italian.

★ Stretch Yourself

Say or write these in Spanish:

1. I hate science because it's complicated but I am interested in maths.
2. My friend and I love drama because it's fun.

Talking About School

School Vocabulary

el / la alumno / a	pupil
el aula (f.)	classroom
el bachillerato elemental	GCSEs (equivalent)
el bachillerato superior	A levels (equivalent)
la cantina	canteen / dining area
la clase	class
el colegio (mixto)	(mixed) school
el curso	course
los deberes	homework
el / la director / ora	headteacher
la escuela	school
el / la estudiante	student
los estudios	studies
el examen	exam
el gimnasio	gym
el instituto	secondary school
el intercambio	school exchange
el laboratorio	laboratory
la lección	lesson
la nota	mark
la página	page
el pasillo	corridor
el patio	playground
la pizarra	blackboard
la pizarra blanca	whiteboard
la pregunta	question
el / la profesor / ora	teacher
la prueba	test
el recreo	break
la respuesta	answer
la sala de profesores	staffroom
el timbre	bell
el trimestre	term
los vestuarios	changing rooms
el vocabulario	vocabulary

¿Cómo es tu instituto?
What's your school like?

Mi instituto es bastante antiguo con unos novecientos alumnos.
My school is quite old with about 900 pupils.

Me gusta mi colegio porque tiene muchas instalaciones buenas como un gimnasio grande, cinco laboratorios modernos, una cantina muy buena y cuatro salas de informática.
I like my school because it has lots of good facilities such as a big gym, five modern labs, a really good canteen and four ICT rooms.

Las clases empiezan a las ocho y media y tenemos un recreo a las once menos cuarto.
Classes start at eight thirty and we have a break at quarter to eleven.

Me gustan muchos de los profesores de mi colegio porque son amables y explican bien.
I like lots of the teachers in my school because they are friendly and they explain things well.

Sin embargo me cae fatal mi profesora de matemáticas porque es muy estricta y no me ayuda mucho.
However, I can't stand my maths teacher because she's very strict and she doesn't help me very much.

Tenemos que llevar uniforme escolar en mi colegio y diría que realmente me gusta. Es una falda gris o pantalones grises, una camisa blanca con un jersey gris y una corbata azul marino.
We have to wear school uniform at my school and I would say that, in truth, I like it. It's a grey skirt or grey trousers, a white shirt with a grey jumper and a navy blue tie.

Creo que es mejor llevar uniforme porque así todo el mundo parece igual y es más elegante.
I think that it's better wearing a uniform because then everyone looks the same and it is smarter.

Build Your Skills: Reflexive Constructions

There are some useful reflexive constructions that, if used successfully, can help to gain higher marks. They are used in Spanish to say 'one / you can...' or 'one does... / you do...'. For example, 'You must wear school uniform' or 'You can use your mobile phone at lunchtime'. These constructions are often used for school rules:

- **Se debe respetar a los demás.**
  You must respect others.
- **No se permite comer chicle en clase.**
  You are not allowed to eat chewing gum in class.
- **Se deben hacer los deberes.**
  You must do your homework.
- **No se debe llevar maquillaje.**
  You must not wear make-up.
- **Solo se puede usar el móvil durante la hora de comer y nunca en clase.**
  You can only use your mobile phone at lunchtime and never in class.

Note that the reflexive verb is followed by the infinitive.

✓ Maximise Your Marks

Avoid losing marks through inaccuracies! See if you can spot the mistakes in the following sentence. There are five of them:

- **Me gusta mi profesores porque son simpático pero las instalaciones en mi colegio es mal.**

It should read:

- **Me gustan mis profesores porque son simpáticos pero las instalaciones en mi colegio son malas.**
  I like (plural, as describing 'teachers') my (plural) teachers because they are kind (plural) but the facilities in my school are (plural) bad (feminine and plural).

Make sure that your adjective agrees with the noun, that you use the correct part of the verb depending on who is doing the action and that impersonal verbs (e.g. **gustar**) are singular or plural depending on the noun.

Education and Work

? Test Yourself

What do these mean in English?

1. **Las clases empiezan a las nueve menos cuarto.**
2. **En mi escuela primaria llevaba uniforme pero ahora no es obligatorio.**

How do you say these in Spanish?

3. I hate my school uniform because it's really ugly.
4. I don't like chemistry because we always have lots of tests.

★ Stretch Yourself

Say or write these in Spanish:

1. At my school, you must always listen to the teacher and respect others.
2. You are not allowed to wear make-up and jewellery.

Pressures and Problems at School

Useful Vocabulary

el acoso escolar	bullying
el apoyo	support
el ataque físico	physical attack
la ayuda	help
el boletín	school report
el comportamiento	behaviour
la conducta	behaviour
los deberes	homework
la disciplina	discipline
el estrés	stress
el examen	exam
el éxito	success
el fracaso	failure
la nota	mark
la presión	pressure
el ruido	noise
la víctima	victim

Useful Verbs

atacar	to attack
callarse	to stop talking / keep quiet
comportarse	to behave
concentrarse	to concentrate
dar igual	to not mind
dar miedo	to scare / frighten
estar estresado / a	to be stressed
estar harto a / de	to be fed up of / with
fastidiar	to annoy
fracasar	to fail
golpear	to hit
gritar	to shout
hacer novillos	to skip / miss school
insultar	to insult
intimidar	to intimidate
justificar	to justify
maltratar	to ill-treat / abuse
molestar	to bother
sacar buenas / malas notas	to get good / bad marks / grades
suspender (un examen)	to fail (an exam)

Academic Pressures

Para mí el mayor problema en el instituto es el estrés de intentar sacar buenas notas todo el tiempo.
For me, the biggest problem at school is the stress of trying to get good grades all of the time.

Estoy harto de pasar todo mi tiempo libre haciendo los deberes. ¡No es justo!
I am fed up with spending all of my free time doing homework. It's not fair!

Para poder ir a la universidad, tengo que aprobar todos mis exámenes con buenas notas.
To be able to go to university, I have to pass all of my exams with good marks.

Mis padres esperan demasiado de mí y como consecuencia me siento estresada.
My parents expect too much of me and as a result I feel stressed.

Tengo miedo de suspender los exámenes y no puedo concentrarme muy bien.
I am scared of failing my exams and I can't concentrate very well.

Los cursos son muy exigentes, aun con el apoyo de los profes.
The courses are very demanding, even with the support of the teachers.

Bullying

En mi colegio, parece que el acoso escolar ocurre raras veces, menos mal.
In my school, it seems that bullying rarely happens, thank goodness.

Hay unos estudiantes mayores en mi cole que intimidan a los más jóvenes y les quitan el dinero. ¡Es insoportable!
There are some older students in my school who intimidate the younger ones and take their money. It's unbearable!

Creo que el problema del acoso escolar existe en todos los institutos hasta cierto punto.
I think that the bullying problem exists in all schools to some degree.

No aguanto a los estudiantes que maltratan a otros, pero yo me callo porque me dan miedo a mí también.
I can't stand students who abuse others, but I keep quiet because they scare me as well.

He empezado a hacer novillos a causa de una estudiante que me insulta todos los días.
I have started to skip school because of a student who insults me every day.

Build Your Skills: More Connectives for Expressing Opinions

You should always be looking to extend your sentences in order to gain more marks. There are many possible connectives and phrases you could use to achieve this. Here are a few more that are good for voicing an opinion:

a causa de	because of
a pesar de	in spite of
aparte de	apart from
claro que...	of course...
dado que	given that
en cuanto a	regarding
en lugar de	instead of
es decir	that is (to say)
sin duda	without doubt

No me gusta mucho ir al instituto a causa de sentirme estresado.
I don't really like going to school because of feeling stressed.

Espero ir a la universidad a pesar de tener que trabajar muchísimo para aprobar los exámenes.
I hope to go university in spite of having to work very hard to pass my exams.

✓ Maximise Your Marks

Generally, nouns ending in **–o** are masculine and nouns ending in **–a** are feminine. However, there are a few exceptions that need to be learnt! Here are some common ones:

Masculine		Feminine	
el clima	climate	**la foto**	photograph
el día	day	**la mano**	hand
el mapa	map	**la moto**	motorbike
el planeta	planet	**la radio**	radio
el problema	problem		
el programa	programme		
el sistema	system		
el tema	topic / theme		

? Test Yourself

What do these mean in English?

1. **El acoso escolar occure bastante en mi colegio.**
2. **Estoy harto de tantos exámenes.**

How do you say these in Spanish?

3. You have to study a lot to pass your exams.
4. I think I am going to fail my English exam.

★ Stretch Yourself

Say or write these in Spanish:

1. The teacher spoke to the student regarding his behaviour.
2. Bullying is, without doubt, one of the biggest problems in schools.

Part-time Jobs and Work Experience

Work Experience Verbs

arreglar	to mend
contactar	to contact
contestar	to answer
diseñar	to design
emplear	to employ
encontrar	to find
escribir	to write
explicar	to explain
ganar	to earn
hacer prácticas	to do work experience
llamar por teléfono	to telephone
obtener	to obtain / get
pagar bien / mal	to pay well / badly
preparar	to prepare / make
probar	to have a go / try
servir	to serve / attend to
trabajar	to work

Talking About Your Work Experience

You can use both the preterite tense and the imperfect tense to talk about what you did on your work experience. Remember that the preterite is for completed actions in the past – 'I worked', 'I helped', 'I did', 'I wrote'; and the imperfect is for things that you used to do, repeated actions or things that you did every day, for example: 'I helped with the filing every day'. The imperfect is also used to describe what someone or something was like: 'My colleagues were very kind'.

Hice mis prácticas laborales en una tienda / oficina / escuela / fábrica / banco.
I did my work experience in a shop / office / school / factory / bank.

Todos los días trabajaba en la oficina. Contestaba llamadas telefónicas y escribía en el ordenador y era bastante interesante, aunque un poco monótono a veces, en mi opinión.
I worked in the office every day. I answered phone calls and typed on the computer and it was quite interesting, although a little monotonous at times, in my opinion.

Me gustaron mucho mis prácticas. Trabajé como profesor en una escuela primaria. Era muy divertido y útil y ahora sé que me gustaría ser profesor en el futuro.
I really liked my work experience. I worked as a teacher in a primary school. It was a lot of fun and useful and now I know that I would like to be a teacher in the future.

Para mí, fue una pérdida de tiempo. No aprendí nada nuevo y no me gustaba estar sin mis amigos todos los días.
For me, it was a waste of time. I didn't learn anything new and I didn't like being without my friends every day.

Tenía que trabajar largas horas desde las ocho hasta las cinco y media y estaba muy cansada por la noche.
I had to work long hours from eight o'clock until half past five and I was very tired in the evening.

✓ Maximise Your Marks

Always look to extend your sentences. This can be done quite simply by adding extra information, such as: where; when; who with; why; what you thought. Think of the five W's to help you to remember this each time you write something!

Hice mis prácticas en julio en un hospital porque me interesa ser médico en el futuro. Mis compañeros eran muy amables y trabajadores y, para mí, fue una experiencia increíble.
I did my work experience in July (when) in a hospital (where), because I'm interested (why) in becoming a doctor in the future. My colleagues (who with) were very friendly and hard-working and, for me (what you thought), it was an incredible experience.

Giving Opinions

Creo que...	I think / believe that...
Diría que...	I would say that...
En mi opinión...	In my opinion...
Me parece que...	It seems to me that...
Opino que...	I think that...
Para mí...	For me...
Pienso que...	I think that...
Por un lado...	On the one hand...
Por otro lado...	On the other hand...
Reconozco que...	I recognise that...
Al fin y al cabo...	At the end of the day...

Part-time Jobs

Los sábados, trabajo en una tienda como dependienta.
On Saturdays, I work in a shop as a sales assistant.

Los fines de semana suelo hacer de canguro y está bien.
On weekends I usually babysit and it's good.

Reparto periódicos todas las mañanas antes de ir al colegio. Me pagan bastante bien.
I deliver newspapers every morning before going to school. I get quite well paid.

Yo lavo coches para los vecinos. ¡Es muy divertido!
I wash cars for the neighbours. It's good fun!

Yo tengo que hacer tareas domésticas para ganar un poco de dinero.
I have to do housework to earn a little money.

Soy camarero en un restaurante en el centro de la ciudad.
I am a waiter in a restaurant in the city centre.

Build Your Skills: Using the Gerund to Talk About Past Experience

Another way to talk about something that you did repeatedly is to use the preterite or imperfect with the gerund (–ing). Remember that this takes the form **–ando** for **–ar** verbs and **–iendo** for **–er** and **–ir** verbs.

Trabajaba como camarero y pasaba mi tiempo sirviendo a los clientes, limpiando las mesas y trabajando en la caja.
I worked as a waiter and I spent my time serving customers, cleaning the tables and working on the till.

Test Yourself

What do these mean in English?

1. **Hice mis prácticas en una tienda.**
2. **Creo que fue una buena experiencia.**

How do you say these in Spanish?

3. I babysit on Saturday nights.
4. In my opinion the experience was very useful.

Stretch Yourself

Say or write this in Spanish:

1. Last year I did my work experience in a bank and it was very interesting. I worked on the till serving the customers, and I would like to work there in the future.

Future Career Plans

Jobs

el / la abogado / a	lawyer
el actor / la actriz	actor / actress
el / la auxiliar de vuelo	air steward / stewardess
la azafata	air hostess
el / la bombero / a	firefighter
el / la cajero / a	cashier
el / la camarero / a	waiter / waitress
el / la cartero / a	postman / postwoman
el / la contable	accountant
el / la dentista	dentist
el / la dependiente / a	sales assistant
el / la electricista	electrician
el / la enfermero / a	nurse
el / la fontanero / a	plumber
el / la ingeniero / a	engineer
el / la mecánico / a	mechanic
el / la médico / a	doctor
el / la peluquero / a	hairdresser
el / la periodista	journalist / reporter
el / la policía	police officer
el / la profesor / ora	teacher
el / la programador / ora	computer programmer
el / la secretario / a	secretary
el / la soldado	soldier
el / la tendero / a	shopkeeper
el / la veterinario / a	vet

When you are talking about what jobs people do, note that the article is not needed:

Soy peluquera.
I am a hairdresser.

Mi padre es médico.
My father is a doctor.

Quiero ser dentista.
I want to be a dentist.

Advantages and Disadvantages of Jobs

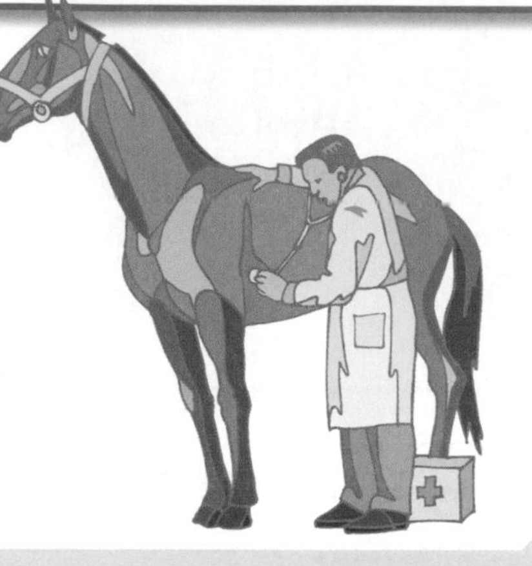

Quiero ser veterinario porque me encantan los animales.
I want to be a vet because I love animals.

Para mí, lo más importante es el sueldo – ¡quiero ganar mucho dinero !
For me, the most important thing is the salary – I want to earn lots of money !

Por un lado, me gustaría ser médico para ayudar a la gente.
On one hand, I'd like to become a doctor to help people.

Por otro lado, las horas son muy largas y eso es difícil.
On the other hand, the hours are very long and that's difficult.

Yo quiero trabajar con niños y tener responsabilidades en mi trabajo.
I want to work with children and have responsibility in my job.

Trabajo bien como parte de un equipo y soy muy trabajador entonces quiero ser bombero.
I work well as part of a team and I am very hard-working so I want to be a firefighter.

No tengo ganas de trabajar en una oficina porque me parece aburrido y monótono.
I don't have any desire to work in an office because it seems boring and monotonous.

Me gustaría ser policía porque quiero hacer algo útil en la vida.
I would like to be a policeman because I'd like to do something useful in life.

Build Your Skills: The Future Tense

The future tense describes what *will happen* or what someone *will do*. It is formed by adding the following set of endings to the infinitive of all regular **–ar**, **–er** and **–ir** verbs:

Hablar To speak
Hablaré I will speak
Hablarás You will speak (**tú**)
Hablará He / She / It will speak; You will speak (**Vd.**)
Hablaremos We will speak
Hablaréis You will speak (**vosotros**)
Hablarán They will speak; You will speak (**Vds.**)

¡Mañana hablaré español!
Tomorrow I will speak Spanish!

Comeremos a las dos y media.
We will eat at half past two.

The following verbs have irregular stems and need to be learnt! However, the endings are regular:

Infinitive	Stem	Future (1st person singular)	
		Spanish	English
decir	dir	diré	I will say
hacer	har-	haré	I will do
poder	podr-	podré	I will be able to
poner	pondr-	pondré	I will put
querer	querr-	querré	I will want
saber	sabr-	sabré	I will know
salir	saldr-	saldré	I will go out
tener	tendr-	tendré	I will have
venir	vendr-	vendré	I will come
		(3rd person singular)	
hay (haber)	har-	habrá	there will be

Note that the future tense is used in constructions with **si** (if) + present + future:

- **Si saco buenas notas, podré ir a la universidad.**
  If I get good grades, I will be able to go to university.
- **Si apruebo mis exámenes, conseguiré un buen trabajo.**
  If I pass my exams, I will get a good job.

Other Ways to Talk About the Future

There are other ways to express the future:

Querer + infinitive:
- **Quiero ir a la universidad el año que viene.**
  I want to go to university next year.

Esperar + infinitive:
- **Espero encontrar un trabajo interesante.**
  I hope to find an interesting job.

Tener la intención de + infinitive:
- **Tengo la intención de seguir estudiando.**
  I intend / plan to continue studying.

Pensar + infinitive:
- **Pienso viajar por el mundo.**
  I'm thinking of travelling around the world.

The conditional tense:
- **Me gustaría trabajar en el extranjero.**
  I would like to work abroad.

Boost Your Memory

Being able to talk about the past, present and future is very important in speaking and writing. Choose ten verbs that can be used in most contexts, such as **hacer** and **ir**, and learn, off by heart, how to use each of them in the present, the past and the future. For example, **hago** (I do), **hice** (I did) and **voy a hacer** (I am going to do); **voy** (I go), **fui** (I went) and **voy a ir** (I am going to go); and so on. This will help to make your speaking more fluent and will save valuable time in your writing as you will only need to double-check the other less common verbs.

Test Yourself

What do these mean in English?

1. **Quiero un trabajo divertido.**
2. **Yo quiero ser periodista.**

How do you say these in Spanish?

3. My friend wants a useful job.
4. I plan to travel with my sister.

Stretch Yourself

Say or write these in Spanish:

1. If I pass my exams, I will continue studying.
2. If I don't go to university, I will work with my dad.

Practice Questions

Complete these exam-style questions to test your skills and understanding. Check your answers on pages 94–95. You may wish to answer these questions on a separate piece of paper.

Reading

1. Read the pupils' comments about school. What are their attitudes towards school subjects?

Estudio muchas asignaturas que me interesan como las ciencias y la informática. El año pasado estudiaba geografía y al final la dejé porque no se me daba muy bien.	Elena
Suelo sacar buenas notas en mis clases, sobre todo en la clase de inglés, que me resulta bastante interesante y fácil.	Paco
Este año estoy estudiando tecnología que es práctica y también informática, que es útil para mí. Lo que pasa es que siempre hay muchos deberes y los odio.	Miguel

By each of the names below, write P (positive), N (negative) or P+N (positive + negative).

a) Elena .... (1) **b)** Paco .... (1) **c)** Miguel .... (1)

2. Read Maria's blog about school life:

Realmente, no me gusta mucho el colegio. No me gustan muchos de los profesores porque son bastante antipáticos y no explican bien. Sin embargo, me cae bien mi profesora de francés porque aunque es estricta, me ayuda mucho. También, creo que los exámenes causan mucho estrés porque hay que trabajar muchísimo y nunca tengo tiempo libre.

Para mí, lo que menos me gusta de mi cole es un grupo de estudiantes que intenta intimidar a otros estudiantes. Pero supongo que el problema del acoso escolar existe en todos los institutos hasta cierto punto y por lo menos los profesores intentan resolverlo.

Tenemos que llevar uniforme, que es muy feo – pantalones marrones y un jersey marrón con camisa blanca, pero por lo menos todos los estudiantes se visten de la misma manera. A mi parecer, esto está mejor porque es más elegante y así todos parecen iguales.

En cuanto a mis amigos – tengo muchos y normalmente charlo con ellos durante el recreo. Es la única oportunidad de estar con ellos que tengo porque por la noche estoy muy ocupada con el trabajo del cole. ¡Qué aburrido!

Now read the sentences below and choose the four that are correct.
Write the letters in the boxes.

☐☐☐☐ (4)

a) María does not like any of her teachers.

b) Her French teacher helps her.

c) The teachers do not do anything about the bullying problem.

d) She thinks bullying is a problem in all schools.

e) María thinks it is good to have a uniform.

f) She likes her uniform.

g) She has a job in the evenings.

h) María spends time with her friends at breaktime.

Speaking

3 You are being interviewed for a school magazine about school life by your Spanish friend. Prepare to talk in Spanish about each of points in the list below.

a) Your favourite subjects and opinions of them

....

....

b) Which subjects you do not like and why

....

....

c) Future study plans

....

....

d) Pressures at school

....

....

e) Your opinion on wearing a school uniform

....

....

f) Your thoughts on your school life in general

....

.... (10)

Writing

4 You are writing in Spanish about work experience and future jobs. You could include information on the following:

- Your work experience: where, when, why, etc.
- What you had to do during work experience
- Your overall opinion of the experience
- Part-time job and opinions
- Your future plans

....

....

....

....

.... (15)

How well did you do?

1–11	Try again	12–20	Getting there	21–26	Good work	27–32	Excellent!

The Subjunctive and the Conditional

Build Your Skills: The Present Subjunctive

The subjunctive is not used very much in English, but it is used quite often in Spanish in certain situations. Some of these are listed below.

The present subjunctive is used when the main verb is in the present tense:

In formal positive commands and all negative commands:

- **¡Coma!** Eat! (**Vd.**)
  ¡No coman! Don't eat! (**Vds.**)
  ¡No comas! Don't eat (**tú**)

When asking / wanting / ordering someone else to do something:

- **Quiero que** vayas **al banco.**
  I want you to go to the bank.

When expressing feelings or emotions about something:

- **Es una pena que no** puedas **ir a la fiesta.**
  It is a pity you can't go to the party.

After **para que** (so that / in order that):

- **Es para que lo** leas**.**
  It's for you to read.

To express doubt or uncertainty:

- **No creo que** vaya**.**
  I don't think that he's going.

After **cuando** when talking about a future event:

- **Cuando** sea **mayor, voy a ser profesor.**
  When I am older, I am going to be a teacher.

To form the present subjunctive, take the first person singular (**yo**) form of the verb, remove the final **o** and add these endings:

	Hablar	Comer	Vivir
(Yo)	hable	coma	viva
(Tú)	hables	comas	vivas
(El / Ella / Vd.)	hable	coma	viva
(Nosotros)	hablemos	comamos	vivamos
(Vosotros)	habléis	comáis	viváis
(Ellos / Ellas / Vds.)	hablen	coman	vivan

Note that apart from the first person singular, the present tense endings of **–ar** verbs are used for the subjunctive of **–er** and **–ir** verbs, while the present tense endings of **–er** verbs are used for the subjunctive of **–ar** verbs.

If a verb has an irregular first person singular in the present indicative (i.e. not subjunctive), the subjunctive will take the same form:

hacer (hago) → haga
decir (digo) → diga

There are also a number of irregular present subjunctive verbs that have to be learnt. For example:

ir → vaya, vayas, vaya, etc.
ser → sea, seas, sea, etc.

Build Your Skills: The Imperfect Subjunctive

The imperfect subjunctive is used in the same situations as the present subjunctive but generally when the main verb is in the preterite or imperfect tense.

The imperfect subjunctive is also used in 'if' clauses in the past:

Si tuviera **mucho dinero, compraría un Ferrari.**
If I had a lot of money, I would buy a Ferrari.

Si ganase **la lotería, viajaría por el mundo.**
If I won the lottery, I would travel the world.

You are not expected to use the imperfect subjunctive at GCSE; however, there may be the occasional imperfect subjunctive used in a reading or listening paper.

Build Your Skills: The Conditional Tense

The conditional tense is used to talk about what someone *would* do or what *would* happen in the future. Using and understanding the conditional will help you to gain higher marks.

Just like the future tense, there is only one set of endings for all verbs, whether **–ar**, **–er** or **–ir**. And as with the future tense, the endings are added to the end of the infinitive:

Hablar	To speak
Hablaría	I would speak
Hablarías	You would speak (**tú**)
Hablaría	He / She / It would speak; You would speak (**Vd.**)
Hablaríamos	We would speak
Hablaríais	You would speak (**vosotros**)
Hablarían	They would speak; You would speak (**Vds.**)
Yo iría mañana.	I would go tomorrow.
Haría muchas cosas.	I would do lots of things.

Verbs that have an irregular stem in the future tense have the same irregular stem in the conditional. As with the future tense, the endings are regular:

decir →	**dir-** →	diría	I would say
hacer →	**har-** →	haría	I would do
poder →	**podr-** →	podría	I would be able to
poner →	**pondr-** →	pondría	I would put
querer →	**querr-** →	querría	I would want
saber →	**sabr-** →	sabría	I would know
salir →	**saldr-** →	saldría	I would go out
tener →	**tendr-** →	tendría	I would have
venir →	**vendr-** →	vendría	I would come
hay (haber) →	**habr-** →	habría	there would be

✓ Maximise Your Marks

The conditional tense endings are the same as the **–er** / **–ir** endings for the imperfect tense. When using the conditional, make sure that you add the ending on to the infinitive (it should have the **–r** at the end). And when listening or reading, look out for that **–r**! This will help you to decide whether the verb is a conditional or an imperfect. For example:

- **Comería** I would eat (conditional)
- **Comía** I ate / used to eat (imperfect)

Build Your Skills: Discussing Complex Issues

There are a number of phrases that lend themselves very well to discussing complex subjects such as the environment and social issues. The following phrases all need the subjunctive:

Es importante que + subjunctive:
- **Es importante que reciclemos la basura.**
  It is important that we recycle rubbish.

Es necesario que + subjunctive:
- **Es necesario que protejamos los bosques y las selvas tropicales.**
  It is necessary for us to protect the forests and tropical rainforests.

Es esencial que + subjunctive:
- **Es esencial que apaguemos las luces y consumamos menos energía.**
  It is essential that we turn off lights and consume less energy.

Es imprescindible que + subjunctive:
- **Es imprescindible que todos hagamos todo lo posible para mejorar la situación.**
  It is essential that we all do as much as possible to improve the situation.

? Test Yourself

What do these mean in English?

1. **Es importante que vayas conmigo.**
2. **¿Podrías salir esta noche?**

How do you say these in Spanish?

3. I want you to speak Spanish!
4. He would drink tea.

★ Stretch Yourself

Say or write these in Spanish:

1. It's a pity that we can't do more.
2. Could you recycle rubbish and turn off lights, please?

Environmental Issues

My Local Environment

Mi pueblo es bonito con pocos edificios y muchas zonas verdes. Realmente no tenemos problemas con el tráfico ni la contaminación.
My town is small with few buildings and lots of green spaces. We don't really have traffic or pollution problems.

Mi ciudad es muy industrial y me preocupa el problema de los desperdicios tóxicos de las fábricas.
My city is industrial and I am worried by the problem of toxic factory waste.

Mi ciudad está sucia y hay muchos atascos a causa de tanto tráfico.
My city is dirty and there are lots of traffic jams because of so much traffic.

Hay contaminación del aire y de los ríos también y no veo ninguna solución.
There is air pollution and also river pollution and I don't see any solution.

Environment Vocabulary

el aire	air
el atasco	traffic jam
la atmósfera	atmosphere
la basura	rubbish
la bolsa de plástico	plastic bag
la campaña	campaign
la capa de ozono	ozone layer
la contaminación	pollution
el daño	damage
los desperdicios	waste
la electricidad	electricity
la energía	energy
el envase	packaging
la gasolina sin plomo	unleaded petrol
el medio ambiente	the environment
el mundo	the world
la naturaleza	nature
el planeta	planet
el problema	problem
los productos químicos	chemicals
el reciclaje	recycling
los recursos naturales	natural resources
el tráfico	traffic
el transporte	transport
el vehículo	vehicle

Useful Verbs

ahorrar	to save
apagar	to turn off
avertir	to warn
ayudar	to help
consumir	to consume
contaminar	to pollute
dañar	to damage
desaparecer	to disappear
desarrollar	to develop
destrozar	to destroy
ducharse	to have a shower
emitir	to emit
encender	to light / turn on
ensuciar	to get dirty
estropear	to spoil
gastar	to use (energy) / spend (money)
malgastar	to waste
mejorar	to improve
preocuparse por	to worry about
producir	to produce
proteger	to protect
reciclar	to recycle
reducir	to reduce
reutilizar	to reuse
salvar	to save
separar la basura	to sort the rubbish
tirar	to throw (away)
usar	to use

Useful Adjectives

ecológico / a	ecological
medioambiental	environmental
mundial	worldwide
peligroso / a	dangerous
químico / a	chemical
recargable	rechargeable
reciclable	recyclable
sucio / a	dirty
tóxico / a	toxic

Looking After the Environment

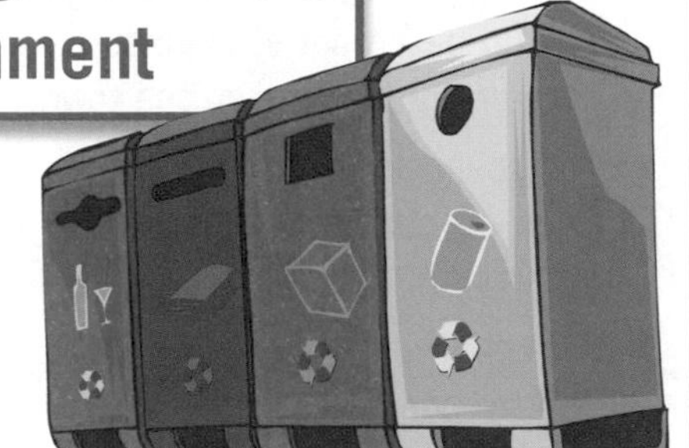

Siempre intento andar en vez de ir en coche.
I always try to walk instead of going by car.

Se debe reciclar todo lo posible como el vidrio, el plástico, el papel y el cartón y meterlo todo en los contenedores de reciclaje.
You should recycle as much as possible such as glass, plastic, paper and cardboard and put it all into the recycling containers.

Para reducir la contaminación, es mejor llevar una bolsa a la compra en vez de pedir una bolsa de plástico.
To reduce pollution, it is better to take a bag to do the shopping, instead of asking for a plastic one.

Es importante apagar las luces y consumir menos energía.
It is important to turn lights off and consume less energy.

Deberíamos usar el transporte público para reducir las emisiones de los coches y para ayudar a prevenir la lluvia ácida.
We should use public transport to reduce car emissions and to help prevent acid rain.

Hay que construir más zonas peatonales y carriles de bicicleta para disminuir la contaminación de ciudades.
They have to build more pedestrianised areas and cycle lanes to reduce the pollution of cities.

Todos tenemos que mejorar la situación.
We all have to improve the situation.

Build Your Skills: The Passive

Verbs can be either active, as in 'He *bought* a coffee', or passive, as in 'It *was made* by him'. In Spanish, the passive is not used as much as it is in English, but you may need to recognise it in a listening or reading exam.

The passive is formed by using the correct part of the verb 'to be', either **ser** or **estar**, and the past participle. Because of the use of **ser** or **estar**, the past participle must agree with the subject. The **ser** / **estar** bit can be in whichever tense is required.

La contaminación del aire es causada por los gases de escape de los coches.
Air pollution is caused by exhaust fumes from cars.

La contaminación será reducida bastante.
Pollution will be reduced considerably.

Boost Your Memory

To help you to revise key vocabulary and phrases for a particular topic, make yourself a spider diagram on a small piece of card. If you get into the habit of doing this for the topics that you struggle with, you can then carry the cards around with you and look at them whenever you get the chance, even if it is just for a few minutes at a time. The more often you glance at your spider diagrams, the easier it will be to remember the vocabulary and phrases!

Test Yourself

What do these mean in English?

1. **Mi ciudad está muy contaminada.**
2. **Se debe ducharse en vez de bañarse.**

How do you say these in Spanish?

3. We should consume less energy.
4. It is better to walk than go by car.

Stretch Yourself

Say or write these in Spanish:

1. Cities are spoilt by so much traffic.
2. The planet will be destroyed by pollution.

Global Issues

Global Vocabulary

agotar	to exhaust
amenazar	to threaten
aumentar	to increase
el aumento	increase
el calentamiento global	global warming
la caza	hunting
los CFCs	CFCs
el combustible fósil	fossil fuel
el consumo	consumption
el crimen	crime
la deforestación	deforestation
la discriminación	discrimination
echar la culpa	to blame
el efecto invernadero	greenhouse effect
la energía eólica	wind power
la energía hidroeléctrica	hydroelectric power
la energía nuclear	nuclear energy
la energía renovable	renewable energy
la energía solar	solar power
la guerra nuclear	nuclear war
inquietante	worrying
inquietar(se)	to worry
la inundación	flood
la marea negra	oil slick
el petrolero	oil tanker
la pobreza	poverty
el recurso	resource
la selva	jungle / forest
la sequía	drought
la sobrepesca	overfishing
la superpoblación	overpopulation
la Tierra	the Earth
el terrorismo	terrorism

When to Use 'Por' and 'Para'

Por and **para** can both mean 'for' but are used in different situations.

Por can mean...

- 'for' or 'in' when talking about time:
  Salí con mis amigos por la noche.
  I went out with my friends at night.
- 'through' when talking about locations:
  Doy un paseo por el pueblo.
  I go for a walk through the town.
- 'in exchange for':
  Le di 40 euros por la chaqueta.
  I gave him 40 euros for the jacket.
- 'on behalf of':
  Lo hizo por ti.
  He did it for you.

Para can mean...

- 'by' or 'in time for':
  Tengo que terminarlo para mañana.
  I have to finish it for tomorrow.
- 'in order to' followed by the infinitive:
  Reciclo mucho para proteger el medio ambiente.
  I recycle a lot to protect the environment.
- 'for' a purpose or recipient:
  Lo compró para su amiga.
  She bought it for her friend.
- 'in my opinion':
  Para mí es muy importante ahorrar agua.
  In my opinion it is really important to save water.

Discussing Global Issues

Lo que más me preocupa es el calentamiento global. Tenemos que usar alternativas a los combustibles fósiles.
What worries me the most is global warming. We have to use alternatives to fossil fuels.

Para mí lo peor es la amenaza del terrorismo que existe en el mundo.
For me, the worst thing is the terrorism threat that exists in the world.

Creo que debemos hacer más para luchar contra la discriminación y el racismo.
I think that we should do more to fight against discrimination and racism.

En mi opinión, la deforestación es un problema muy serio.
In my opinion, deforestation is a very serious problem.

Debemos crear un mundo mejor para nuestros hijos.
We should create a better world for our children.

Build Your Skills: Relative Pronouns

Relative pronouns relate back to something or someone previously mentioned in the sentence. For example:

- **La chica que habla es mi hermana.**
  The girl who is talking is my sister.

The most common relative pronoun in Spanish is **que**, which translates as 'that', 'which' or 'who'. This is sometimes left out in English, but in Spanish it must be used.

- **El libro que lee es mío.**
  The book (that) he is reading is mine.

The pronouns **el que**, **la que**, **los que**, **las que** are usually used after prepositions:

- **El asunto del que hablamos, es muy serio.**
  The subject about which we are talking, is very serious.

In the above example, the preposition **de** is joined with **el** to make **del**.

El cual, **la cual**, **los cuales** and **las cuales** are used in the same way, but tend to be more formal:

- **El asunto del cual hablamos, es muy serio.**

Lo que refers to an abstract concept and often translates as 'what':

- **Lo que más me molesta es que muchas personas no tienen acceso al agua potable**.
  What bothers me the most is that many people do not have access to drinking water.

Quien and **quienes** (plural) are only used when referring to people and usually after a preposition:

- **El profesor con quien hablaba es mi profesor de música.**
  The teacher that I was talking to (to whom I was talking) is my music teacher.

✓ Maximise Your Marks

When reading or listening, remember to use the context to help you work out meaning. These topics are quite emotive and there will often be indications of positive/negative emotions etc. to aid comprehension. Also, look out for cognates, e.g. **la deforestación**, which help understanding.

Verbs with the Infinitive and a Preposition

Some verbs can be followed by a simple infinitive, for example, **espero mejorar...** (I hope to improve...). However, other verbs require a preposition before the infinitive:

tratar de	to try to
acordarse de	to remember to
olvidarse de	to forget to
acabar de	to have just
terminar de	to finish...-ing
aprender a	to learn to
ayudar a	to help to
empezar a	to begin to
invitar a	to invite to
forzar a	to force to

Trato de ducharme en vez de bañarme.
I try to have a shower instead of a bath.

? Test Yourself

What do these mean in English?

1. **Hay que buscar energía renovable.**
2. **Para mí, la sequía es muy preocupante.**

How do you say these in Spanish?

3. What worries me the most is the greenhouse effect.
4. We have to learn to create a better world.

★ Stretch Yourself

Say or write these in Spanish:

1. The person that worries me the most is my sister.
2. What we have to do is improve the situation.

The Wider World

Social Issues

Issues Affecting Society

la anorexia	anorexia
la bulimia	bulimia
el cáncer	cancer
la delincuencia	delinquency / crime
los derechos humanos	human rights
el desempleo	unemployment
la exclusión social	social exclusion
los gamberros	hooligans
la igualdad	equality
la indignidad	indignity
los inmigrantes	immigrants
los minusválidos	disabled people
el paro	unemployment
el racismo	racism
el respeto	respect
la responsabilidad	responsibility
el sexismo	sexism
el SIDA	AIDS
los sin techo / hogar	homeless people
el / la voluntario / a	volunteer

En todos los países del mundo hay gente que vive en la calle. Es muy preocupante.
In every country in the world there are people living in the street. It is very worrying.

Tenemos que luchar contra el racismo porque creo que los derechos humanos son muy importantes.
We have to fight against racism because I believe that human rights are very important.

El desempleo es algo que afecta a gente por todo el país.
Unemployment is something that affects people all over the country.

Hay que educar a la gente para cambiar su manera de pensar.
We must educate people to change their way of thinking.

Siempre hay que estar muy flaca como las modelos en la tele y por eso trato de no comer mucho.
You always have to be thin like the models on TV and because of that I try not to eat much.

Tackling the Issue

adelgazar	to lose weight
asegurar	to assure
beneficiar	to benefit
castigar	to punish
compadecer	to feel sorry for
contribuir	to contribute
conversar	to relate / report
cuidar	to look after
dar las gracias	to thank
decidir	to decide
detener	to arrest
disculpar(se)	to apologise
discutir	to discuss
dudar	to doubt
educar	to educate
equivocarse	to be wrong / make a mistake
estar en paro	to be unemployed
fiarse de	to trust
formar parte de	to be part of
hacerse voluntario / a	to become a volunteer
lograr	to achieve / manage to
maltratar	to mistreat
ofender	to offend
ofenderse	to be offended
quejarse	to complain

Voy a hacerme voluntario para ayudar a los demás.
I'm going to become a volunteer to help others.

En el pasado era un poco egoísta pero ahora trato de dar mi tiempo a otros.
In the past I was a bit selfish, but now I try to give my time to others.

Si nos juntamos y trabajamos juntos, lograremos mejorar nuestra sociedad.
If we all come together and work together, we will manage to improve our society.

Hay un problema con la delincuencia en muchos sitios y tenemos que crear oportunidades para que esto se acabe.
There is a problem of delinquency in many places and we have to create opportunities so that it stops.

Build Your Skills: Possessive Adjectives

Possessive adjectives show who something belongs to. They can be a little confusing in Spanish as they have to agree with the noun that they are describing and *not* with the person who possesses the thing. Marks for accuracy can be affected if they are not used correctly in the exam. In Spanish, possessive adjectives come in front of the noun.

Here are the possessive adjectives in full:

Masculine Singular	Feminine Singular	Masculine Plural	Feminine Plural	English
mi	**mi**	**mis**	**mis**	my
tu	**tu**	**tus**	**tus**	your (**tú**)
su	**su**	**sus**	**sus**	his / her / its / your (**Vd.**)
nuestro	**nuestra**	**nuestros**	**nuestras**	our
vuestro	**vuestra**	**vuestros**	**vuestras**	your (**vosotros**)
su	**su**	**sus**	**sus**	their / your (**Vds.**)

La igualdad es su derecho.
Equality is his right.

Son nuestros problemas.
They are our problems.

Es el futuro de tu hijo.
It's your son's future.

Es nuestra responsabilidad de luchar contra la desigualidad.
It's our responsibility to fight against inequality.

✓ Maximise Your Marks

There are a number of 'false friends' in Spanish – words that look very similar to an English word, but have a different meaning. Make sure you learn them so as to avoid confusion. For example, **sensible** means 'sensitive' and *not* 'sensible', **simpático** means 'kind' / 'nice' and *not* 'sympathetic'.

Here are a few others:

embarazada pregnant
estar constipado to have a cold
vago lazy

? Test Yourself

What do these mean in English?

1. **Quiero hacerme voluntario.**
2. **Estoy intentando contribuir más.**

How do you say these in Spanish?

3. There is a lack of respect in society.
4. We must educate people.

★ Stretch Yourself

Say or write these in Spanish:

1. It is our responsibility to improve society.
2. Everyone has their human rights.

Technology

Technology Vocabulary and Verbs

la banda ancha	broadband
el chat	chat room
el cibercafé	Internet café
el correo electrónico / email	email
un disco compacto	CD
el disco duro	hard drive
la impresora	printer
el iPod	iPod
el lector / reproductor de MP3	MP3 player
el mensaje (de texto)	text message
el monitor	monitor
el (teléfono) móvil	mobile phone
el (ordenador) portátil	laptop
el ratón	mouse
la red	Internet / web
el sitio web	website
el teclado	keyboard
chatear	to chat
conectar(se)	to connect
descargar	to download
desconectar	to disconnect
mandar	to send
navegar por Internet	to surf the net

Advantages and Disadvantages of Technology

Me gusta usar el portátil por la noche para hacer mis deberes y luego para chatear con mis amigos. Lo encuentro muy práctico.
I like to use my laptop in the evening to do my homework and then to chat with my friends. I find it very convenient.

¡Yo no podría vivir sin el móvil! Lo uso todo el día. Mando mensajes a mis amigos, navego por Internet y, más tarde, llamo a gente. ¡Está siempre conmigo!
I couldn't live without my mobile! I use it all day long. I send messages to friends, I surf the Internet and, later on, I ring people. It's always with me!

A mí me encanta descargar música del Internet y es más fácil que ir a las tiendas y comprar un disco compacto.
I love to download music from the Internet and it's easier than going to the shops and buying a CD.

Los chats pueden ser peligrosos porque los criminales pueden usarlos. ¡Hay que tener cuidado!
Chat rooms can be dangerous because criminals can use them. You have to be careful!

Cuando voy de paseo con mi perro, me gusta llevar mi iPod para escuchar música. Así el tiempo pasa más rápido.
When I go for a walk with my dog, I like to take my iPod to listen to music. That way, the time passes more quickly.

En mi opinión hacer compras por Internet es más barato que ir de compras en la ciudad.
In my opinion shopping on the Internet is cheaper than going shopping in the city.

Me molesta mucho que por la noche mis amigos nunca quieren salir porque siempre quieren usar Facebook o hablar en los chats. Para mí es muy aburrido.
It bothers me a lot that at night my friends don't want to go out because they always want to go on Facebook or talk on chat rooms. It's really boring for me.

Build Your Skills: Possessive Pronouns

To ask who something belongs to, you can use the question **¿De quién es?** You would answer that with one of the following possessive pronouns, making sure that it agreed with the noun that it replaces:

¿De quién es este móvil? ¿Es tuyo?
Whose is this mobile phone? Is it yours?

Sí, es mío.
Yes, it's mine.

¿De quien es ese iPod? ¿Es suyo?
Whose is that iPod? Is it his?

No, es tuyo.
No, it's yours.

When the possessive pronoun does not follow the verb **ser**, you need to add the relevant definite article (**el / le / los / las**) in front of the pronoun.

- **En cuanto a los móviles, el mío es más nuevo que el tuyo.**
  As for mobiles, mine is newer than yours.
- **Mi portátil tiene un monitor muy pequeño. ¿Y el tuyo?**
  My laptop has a very small monitor. And yours?
- **Mi banda ancha sale más barata que la tuya.**
  My broadband works out cheaper than yours.
- **Mi amigo acaba de comprar una impresora nueva. Es mucho más moderna que la mía.**
- My friend has just bought a new printer. It's much more modern than mine.

Masculine Singular	Feminine Singular	Masculine Plural	Feminine Plural	English
mío	**mía**	**míos**	**mías**	mine
tuyo	**tuya**	**tuyos**	**tuyas**	yours (**tú**)
suyo	**suya**	**suyos**	**suyas**	his / hers / its / yours (**Vd.**)
nuestro	**nuestra**	**nuestros**	**nuestras**	ours
vuestro	**vuestra**	**vuestros**	**vuestras**	yours (**vosotros**)
suyo	**suya**	**suyos**	**suyas**	theirs / yours (**Vds.**)

✓ Maximise Your Marks

In speaking assessments, it is good to have a few 'fillers' to use to give you a chance to think of what you are going to say. Phrases such as **pues...** (well...); **a ver...** (let's see...); **un momento...** (one moment...); **bueno...** (well...) are all easy to remember and will give you a moment to collect your thoughts!

? Test Yourself

What do these mean in English?

1. **Los correos electrónicos son más rápidos.**
2. **Creo que chatear por Internet es muy peligroso.**

How do you say these in Spanish?

3. I surf the net every day.
4. I use my mobile to send texts.

★ Stretch Yourself

Say or write these in Spanish:

1. Is that your laptop? No, it's his.
2. Mobile phones? Hers is really good. Better than ours.

Practice Questions

Complete these exam-style questions to test your skills and understanding. Check your answers on page 95. You may wish to answer these questions on a separate piece of paper.

Reading

1 Read the following letter and response from a newspaper and then answer the questions.

Problema
Soy una chica de 15 años y me preocupa mucho el medioambiente. Pero en mi casa no hacemos nada ni para reciclar ni reutilizar nada. Encima, siempre dejamos encendidas las luces, no separamos la basura para reciclarla y mi madre se baña todas las noches. Yo intento ir al cole a pie pero mi padre prefiere llevarme en coche porque dice que es más cómodo para mí. ¡Ayúdeme por favor! Isabel

Respuesta
Yo creo que tienes que hablar con tus padres para explicarles el problema. Explícales lo importante que es para ti y que quieres que te ayuden. Entonces tu madre puede ducharse para ahorrar agua, todos podéis meter la basura en los contenedores correctos y cuando os vais de compras, debéis llevar bolsas para no tener que pedir bolsas de plástico. María

a) Name five things that show that Isabel's family are not environmentally friendly.

....

....

.... (5)

b) Name two things that María suggests Isabel tells her family.

.... (2)

c) What two things does María suggest they could all do to help?

.... (2)

2 Read the headlines below and match them to the statements that follow.

A **ES ESENCIAL QUE LUCHEMOS CONTRA EL EFECTO INVERNADERO.**

B ***El 80% de las personas sin hogar son hombres.***

C **He pasado el mejor año de mi vida ayudando a mucha gente – y ¡no me pagan nada!**

D **67 MUERTOS Y 29 HERIDOS EN LA CAPITAL, DESPUÉS DE UNA NOCHE HORROROSA.**

E **Somos todos iguales y tenemos el derecho a vivir juntos con dignidad.**

F **No quiero salir por la noche por si me intimidan o abusan de mí.**

a) Life as a volunteer ☐ (1) **b)** Terrorism ☐ (1)

c) Against discrimination ☐ (1) **d)** Global warming ☐ (1)

e) Crime and delinquency ☐ (1) **f)** Homelessness ☐ (1)

Speaking

3 You are being interviewed for a school magazine about the environment by your Spanish friend. Prepare to speak in Spanish about each of the points in the list below.

a) Your thoughts on the environment

....

b) What you do at home to help the environment

....

c) What else you would like to do in the future

....

d) What you see as environmental problems in your local area

....

e) What you think should be done locally

....

f) Your opinions on the environment on a global scale

.... (10)

Writing

4 You are writing about the advantages and disadvantages of technology for a Spanish IT magazine. Write the article in Spanish and include information on the following:

- How often you use ICT and where you use it
- What you use ICT for in general and why
- What your parents used to use to do the same tasks
- What you see as the advantages of ICT
- What you see as the disadvantages of ICT
- How you think people will use ICT in the future

....

....

....

....

....

....

....

.... (15)

How well did you do?

1–13	Try again	14–23	Getting there	24–32	Good work	33–40	Excellent!

Word Bank

Here are some additional items of vocabulary for each of the five topic areas. These words are only likely to come up in higher tier listening and reading questions.

Home Life

abrir	to unlock
el aire acondicionado	air-conditioning
alabar	to praise
la altura	height
la amistad	friendship
los ancianos	senior citizens
atraer	to attract
burlar(se) de	to joke (mock)
célebre	famous
la cerradura	lock (of door)
el chubasquero	cagoule
la cólera	anger
cuidado(so)	care(ful)
curioso	odd / strange / curious
decepcionado	disappointed
decepcionar	to disappoint
doblar	to fold
eficaz	efficient
engordar	to put on weight
enojar(se)	to annoy (to get annoyed)
estar avergonzado	to be ashamed
estar contento (de)	to be pleased (about)
llamar	to knock (door)
el mando a distancia	remote control
maquillarse	to put on make-up
el matrimonio	married couple / marriage
medir	to measure
minúsculo	tiny
un montón	pile / load (informal)
el placer	pleasure
poseer	to possess
el prefijo	dialling code
la preocupación	worry / concern
querer decir	to mean
regar	to water (plants)
semejante	similar
el sentido	sense
el sentimiento	feeling
el siglo	century
silencioso	silent
la silla de ruedas	wheelchair
la silla plegable	folding chair
la sonrisa	smile
sordo	deaf
suave	gentle / delicate / mild
el / la suegro / a	father- / mother-in-law
el sueño	dream
terco	obstinate
el triunfo	triumph
valioso	valuable
el vapor	steam
la vela	candle

Home and Away

abrocharse el cinturón	to fasten seat belt
el acantilado	cliff
adelantar	to overtake
anular	to cancel / validate a ticket
averiado	out of order (not working)
cargar	to load
el cinturón de seguridad	seat belt / safety belt
el círculo	circle
concurrido	busy / crowded
la cuesta	slope
cultivar	to grow (plants)
desembarcar	to disembark
el desvío	diversion
encontrarse	to be situated
fluir	to flow
frenar	to brake
la hora punta	rush hour
el límite de velocidad	speed limit
la matrícula	registration number
la orilla	bank (lake / river)
rodeado de	surrounded by
situar(se)	to be situated
la tintorería	dry cleaners
tocar el claxón	to hoot
el volante	steering wheel
la cabaña	hut
calentar(se)	to heat; get warm
echar de menos	to miss (person / place)
la escarcha	frost
fundirse	to melt
granizar	to hail
el granizo	hail
la guía	guide book
húmedo	damp
los intervalos de sol	bright periods
inundar	to flood
la isla	island
la luna	moon
el relámpago	lightning
la sombra	shade
soplar	to blow
tibio	lukewarm
tronar	to thunder
el trueno	thunder
variable	changeable

Free Time and Leisure

acompañar	to accompany
el acontecimiento	event / occurrence
asistir a	to be present at
la cita	appointment / engagement
el descuento	discount / reduction
el día festivo	public holiday
el disfraz	disguise / fancy dress
disfrutar	to enjoy
la diversión	entertainment
entusiasmar	to excite / enthuse
la feria	fair (show)
grabar	to record / burn (CDs)
el martes de Carnaval	Shrove Tuesday
mirar escaparates	to window-shop
el noviazgo	engagement
el ocio	leisure
sorprender	to surprise
el suceso	event
tener lugar	to take place
la ambulancia	ambulance
el asesinato	murder
el atraco	mugging / hold-up
el aviso	warning
la ayuda	help
la cárcel	prison
el choque	collision
el crimen	crime
el daño	damage
el desastre	disaster
la escena	scene
el golpe	blow
el grito	shout
el incendio	fire
la inundación	flood
la multa	fine
el ladrón / la ladrona	thief
el pinchazo	puncture
la recompensa	reward
el robo	robbery
la sangre	blood
el testigo	witness
la tragedia	tragedy
la víctima	victim

Education and Work

las ciencias sociales	social science(s)
la cifra	figure (statistics)
contar	to count / tell (story)
la encuesta	survey / enquiry
la enseñanza	teaching
entregar	to hand in
fracasar	to fail
hábil	clever / skilful
la habilidad	skill / ability
repetir el curso	to repeat the year
el sondeo	survey
subrayar	to underline
la sugerencia	suggestion
el tablón (de anuncios)	notice board
aconsejar	to advise
cobrar	to earn (money)
despedir	to sack
el / la empleado / a	employee
la entrevista	interview
tener éxito	to succeed
la experiencia laboral	work experience
la ficha	file (papers)
el gerente	manager
la licenciatura	degree (university)
ocuparse de	to deal with
la oficina de empleo	job-centre
la orientación profesional	careers advice
planear	to plan
ponerse a + inf.	to start / begin to do something
solicitar	to apply (for a job)
tener en cuenta	to bear in mind / take into account
la traducción	translation
traducir	to translate

The Wider World

ahogarse	to drown
atreverse (a)	to dare (to)
atropellar	to run over / knock down
ceder	to give in / give up
las cerillas	matches
cesar	to cease / stop
chocar con	to collide with
el contrario	opposite
dejar caer	to drop
descuidado	careless / neglected
detener	to arrest
disculpar(se)	to apologise
engañar	to deceive / cheat
envenenar	to poison
esconder	to hide
espantoso	appalling
jurar	to swear
la lágrima	tear (weeping)
la ley	law (justice)
llorar	to cry
malentendido	misunderstanding (confusion)
mentir	to tell lies
la mentira	lie
morder	to bite
la paz	peace
pegar	to stick / hit
perderse	to get lost
la pesadilla	nightmare
prever	to foresee
la reina	queen
el rey	king
la seguridad	safety / security
sospechar	to suspect
tirar	to shoot
la urgencia (llamada de)	emergency (call)

Answers

Guidance for the Speaking and Writing Answers to the Practice Questions

Speaking

Marks will be awarded as follows:

9–10 Marks
Very Good
This means you have covered all the points and given detailed answers, including plenty of relevant information. You have spoken clearly, and have included opinions and reasons for your opinions. You have used some longer sentences and you have used more than one tense.

7–8 Marks
Good
This means you have covered all the points but one of the points may not be as detailed as the others. You have given quite a lot of information clearly, and have included some opinions and reasons. You have used some longer sentences and you have used more than one tense.

5–6 Marks
Sufficient
This means you might not have covered one or two of the points but what you have said conveys some information and there are opinions expressed. Most of your sentences are quite short and your answer may not show much evidence of different tenses.

3–4 Marks
Limited
This means that you have spoken in brief sentences and included some simple opinions but your answer lacks detail and you have missed out some of the information you were asked to give. Your sentences are short and you have used only one tense.

1–2 Marks
Poor
This means that you could not really answer the question and that you gave very little information and expressed no opinions. All your sentences are short and in the same tense.

Writing

Marks will be awarded as follows:

13–15 Marks
Very Good
This means you have covered all the bullet points and given a detailed answer, including plenty of relevant information. You have written clearly, and have included opinions and reasons for your opinions. You have set out your work in a logical and clear structure. You have used some longer sentences and you have used more than one tense.

10–12 Marks
Good
This means you have covered all the bullet points but one of the points may not be as detailed as the others. You have given quite a lot of information clearly, and have included some opinions and reasons. There are some longer sentences and you have used more than one tense.

7–9 Marks
Sufficient
This means you might not have covered one or two of the bullet points but what you have written conveys some information and there are opinions expressed. Most of your sentences are quite short and your answer may not show much evidence of different tenses.

4–6 Marks
Limited
This means that you have written some brief sentences and included some simple opinions but your answer lacks detail and you have missed out some of the information you were asked to give. Your sentences are short and you have used only one tense.

1–3 Marks
Poor
This means that you could not really answer the question and that you have given very little information and expressed no opinions. All your sentences are short and in the same tense.

Note that while sample answers are provided for most of the speaking and writing sections of the practice questions, not every bullet point is covered.

Pages 6–7 Basic Phrases and Expressions
Test Yourself Answers
1 It is twenty past four.
2 In January it is cold.
3 Son las diez y media.
4 Está lloviendo y hay niebla.

Stretch Yourself Answers
1 Llego el veintiuno de enero.
2 14 de octubre de dos mil once

Home Life

Pages 8–9 Personal Information
Test Yourself Answers
1 I think that we are friendly.
2 My birthday is the 22nd of April.
3 Vivo en Glasgow desde hace dos años.
4 En mi opinión son simpáticos.

Stretch Yourself Answers
1 Diría que soy trabajador pero mi madre dice que soy perezoso.
2 En mi opinión mi profesor de inglés es gracioso pero mi amigo dice que es estricto.

Pages 10–11 Family and Friends
Test Yourself Answers
1 Mi prima tiene el pelo castaño y rizado.
2 Soy bastante alto y tengo los ojos azules.
3 Mi tía es de estatura mediana. Mi abuelo es bastante bajo.
4 Mi abuela es muy delgada. Mi hijo tiene pecas.

Stretch Yourself Answers
1 En mi opinión mi hermano es bastante guapo porque es alto con los ojos verdes y el pelo moreno.
2 Mi mejor amiga tiene quince años. Tiene los ojos marrones y el pelo rubio y nació en mil novecientos noventa y seis.

Pages 12–13 The Present Tense
Test Yourself Answers
1 My sister lives in a big house.
2 We eat at half past one.
3 Hablo español, hablas francés.
4 Salgo con mis amigos.

Stretch Yourself Answers
1 Quiero saber la fecha.
2 Nunca como en el comedor.

Pages 14–15 Relationships
Test Yourself Answers
1 My grandmother is widowed.
2 My father is really fun.
3 Mi hermano está soltero.
4 Mis abuelos están separados.

Stretch Yourself Answers
1 Me llevo bien con mi hermano pero mi hermana es enojosa a veces.
2 No me llevo bien con mi hermano porque es egoísta.

Page 16–17 Radical-changing Verbs
Test Yourself Answers
1 I enjoy myself a lot / have fun with my friends.
2 How many hours do you sleep?
3 ¿Qué piensas?
4 Siguen las instrucciones.

Stretch Yourself Answers
1 ¿A qué hora te vistes normalmente los días escolares?
2 La casa cuesta mucho dinero.

Page 18–19 Daily Routine
Test Yourself Answers
1 I comb my hair. You get dressed. We wake up.
2 He / She wakes up. You get washed. You go to bed.
3 Normalmente me visto a las siete y media.
4 Mi hermano se levanta a las ocho menos cuarto.

Stretch Yourself Answers
1 Normalmente me acuesto a las once pero hoy voy a acostarme a las diez.
2 Todos los días me levanto a las siete pero los sábados me quedo en la cama hasta las nueve.

Pages 20–21 House and Home
Test Yourself Answers
1 The lamp is on (top of) the table.
2 The garage is at the end of the garden.
3 Diría que mi dormitorio es bastante grande.
4 En mi casa hay seis habitaciones.

Stretch Yourself Answers
1 Mi dormitorio es mi habitación preferida porque es acogedor y los muebles son bonitos.
2 El jardín está detrás de la casa.

Pages 22–23 Helping at Home
Test Yourself Answers
1 I never wash the car.
2 My sister never does the cooking.
3 Lavo la ropa.
4 Pasea al perro.

Stretch Yourself Answers
1 Hoy tengo que fregar los platos y mi hermano tiene que ordenar su dormitorio.
2 Ayer tuve que pasear al perro.

Pages 24–25 Answers to Practice Questions
(See the guidance on page 90.)
1 **a)** perezoso
b) simpático
c) listo
d) reservado
e) deportivo
f) divertido
2 Laura P+N
Manolo P
Nuria N
3 Example answers:
a) En mi familia somos cuatro, mis padres, mi hermana y yo. Mi madre es médica y trabaja mucho, mi padre es mecánico y nació en mil novecientos sesenta. Mi hermana se llama Sara y es baja con el pelo largo y lleva gafas.
There are four of us in my family, my parents, my sister and myself. My mum is a doctor and she works a lot, my dad is a mechanic and he was born in 1960. My sister is called Sara and she is short with long hair and she wears glasses.
c) Mi casa es bastante pequeña con tres dormitorios. Comparto mi dormitorio con mi hermana pero me da igual. Tenemos un jardín grande y al lado del jardín hay el garaje. El año próximo vamos a comprar una casa más grande.
My house is quite small with three bedrooms. I share a bedroom with my sister but I don't mind. We have a big garden and next to the garden there is the garage. Next year we are going to buy a bigger house.
e) Ayer por la tarde, hice mis deberes de historia y de matemáticas y entonces vi la tele un poco. Después de cenar, escribí un correo electronico a mi amiga española y luego escuché música con mi hermana. Me bañé a las nueve y me acosté a las diez porque estaba muy cansada.
Yesterday afternoon, I did my history and maths homework and then I watched the TV a bit. After dinner, I wrote an email to my Spanish friend and then I listened to music with my sister. I had a bath at nine o'clock and I went to bed at ten because I was very tired.
f) Tengo que ayudar en casa. Normalmente yo pongo y quito la mesa y mi hermana frega los platos. Tengo que hacer la cama y a menudo ayudo a mi madre a preparar la comida. Hoy me toca barrer el suelo también, que es aburrido, pero realmente no me importa.
I have to help at home. Usually I set and clear the table and my sister washes the dishes. I have to make my bed and I often help my mum to do the cooking. Today it's my turn to sweep up as well, which is boring, but I don't mind really.
4 Example answer:
Me llamo Virginia y soy cantante. Soy bastante alta con el pelo corto y moreno y tengo veintisiete años. Soy cantante desde hace un año y medio y, bueno, ahora soy muy famosa.

Por la mañana me levanto bastante temprano, sobre las seis y media, y paseo al perro por la playa. Luego desayuno y voy a mi clase de aerobic. Después, tengo que practicar con mi grupo y por la tarde tenemos que hacer entrevistas y practicar más.

Antes mi casa era muy pequeña con sólo dos dormitorios y un cuarto de baño y no había jardín. Ahora tengo una casa mucho más grande con seis dormitorios y cuatro cuartos de baño. También hay una piscina, que me encanta.

Diría que mis padres me han influido mucho durante mi vida porque siempre me han escuchado y siempre he podido hablarles de cualquier cosa. Creo que es muy importante tener a alguién con quien puedes hablar de tus esperanzas y tus preocupaciones.

I am called Virginia and I am a singer. I am quite tall with short, dark hair and I am 27 years old. I have been a singer for one and a half years and, well, I am very famous now!

In the morning I get up quite early, at about half past six, and I take my dog for a walk along the beach. Then I have breakfast and go to my aerobic class. Afterwards I have to practise with my band and in the afternoon we have to do interviews and practise some more.

Before, my house was very small with only two bedrooms and one bathroom and there wasn't a garden. Now I have a much bigger house with six bedrooms and four bathrooms. There is also a swimming pool, that I love.

I would say that my parents have influenced me a lot in my life because they have always listened to me and I have always been able to talk to them about anything. I think it is really important to have someone to whom you can talk about your hopes and your worries.

Home and Away

Pages 26–27 Local Area
Test Yourself Answers
1 I live in an industrial city.
2 My town / village is very noisy.
3 Mi pueblo puede ser bastante aislado.
4 Se puede visitar la catedral.

Stretch Yourself Answers
1 Lo mejor de mi pueblo es que se puede hacer muchas cosas divertidas con los amigos.
2 Lo bueno de mi pueblo es el nuevo centro comercial.

Page 28–29 Getting Around
Test Yourself Answers
1 Go straight on.
2 Cross the road.
3 Está a la izquierda.
4 Siga todo recto hasta los semáforos.

Stretch Yourself Answers
1 Por favor, ¡levántate temprano mañana!
2 ¡No bajéis la carretera!

Page 30–31 Holiday Plans and Preparations
Test Yourself Answers
1 I usually go on holiday to France.
2 I'm going to sunbathe / get a suntan on the beach.
3 El año que viene voy a ir a España.
4 Vamos a sacar muchas fotos.

Stretch Yourself Answers
1 Antes de hacer turismo, voy a nadar en la piscina y tomar el sol un poco.
2 Después de comer en el restaurante, voy a dar un paseo.

Page 32–33 Accommodation
Test Yourself Answers
1 I would like a single room.
2 There are four adults and three children.
3 Quisiera una habitación doble, por favor.
4 ¿Hay sitio para una caravana y un coche?

Stretch Yourself Answers
1 La luz en el dormitorio no funciona y no hay toallas en el cuarto de baño.
2 He perdido mis llaves y mi monedero.

Page 34–35 The Preterite Tense
Test Yourself Answers
1 They drank a bottle of water.
2 I went out with my friends last night.
3 llegué visité vi
4 comiste bebimos dormiste

Stretch Yourself Answers
1 El año pasado mi padre leyó la carta en francés.
2 Saqué muchas fotos del hotel.

Page 36–37 Holiday Activities
Test Yourself Answers
1 I always dance at the disco.
2 It was good weather, sunny and very hot.
3 Vi muchos lugares de interés.
4 Voy a volver el verano próximo.

Stretch Yourself Answer
1 Primero fuimos a la playa mientras que mis padres se relajaron en el balcón. Luego visitamos un pueblo bonito y finalmente comí en un restaurante con mi familia y mis amigos.

Pages 38–39 Answers to Practice Questions
(See the guidance on page 90.)
1 **a)** C **b)** F **c)** E **d)** A **e)** D **f)** B
2 **a)** 1 month
b) In her own apartment on the coast
c) See her friends / Spend all day at the beach
d) Swim / Practise water sports
e) To a village in the mountains
f) Took photos / Bought souvenirs for grandparents
g) In a restaurant near their apartment
3 Example answers:
b) La ciudad es grande y tiene de todo. Hay tres cines grandes y dos polideportivos que son muy buenos. Los dos tienen una piscina y voy todas las semanas con mis amigos. Hay una gran variedad de tiendas y centro comerciales y por la noche se puede cenar en muchos restaurantes.
The city is big and has everything. There are three big cinemas and two sports centres. Both have a swimming pool and I go every week with my friends. There is a large variety of shops and shopping centres and at night you can eat in many restaurants.
c) Si te gusta ir de compras hay una mezcla de tiendas de diseño y almacenes grandes. Se puede visitar unos lugares de interés también como el barrio antiguo con sus calles estrechas y la catedral antigua, o pasear por el río y ver los barcos.
If you like shopping, there is a mixture of designer shops and department stores. You can visit some interesting places as well, such as the old neighbourhood with its narrow streets and the old cathedral, or go for a walk by the river and see the boats.
d) En el pasado era bastante industrial porque hay un puerto aquí y había muchos barcos viniendo de todo el mundo. Ahora no hay tanta industria. Tampoco había muchas instalaciones para los jóvenes pero lo han renovado y ahora es mucho mejor.
In the past, it was quite industrial because there is a port here, and there used to be lots of boats coming from all over the world. Now there isn't as much industry. Neither were there many facilities for young people but it has been renovated and now it's much better.
f) Yo uso mucho el transporte público, que está muy bien aquí. Puedo coger el autobús desde mi casa para ir al centro y, en la ciudad, hay una red de metro. Sé que tenemos suerte aquí porque en muchos sitios el transporte no es muy bueno. Lo único es que es bastante caro.
I use public transport, which is very good here. I can catch the bus from my house to the city centre and, in the city, there is an underground train network. I know that we are lucky here because in lots of places transport isn't very good. The only thing is that it's quite expensive.
4 Example answer:
El año pasado fui de vacaciones a Italia con mi familia y mi mejor amigo. Viajamos en coche y el viaje duró tres días en total. Pasamos por el túnel de la Mancha y luego viajamos por Francia hasta los Alpes. Llegamos a Italia muy cansados pero muy contentos.

Nos quedamos en un camping bonito cerca de un lago enorme. Había buenas instalaciones y el paisaje era precioso. Lo que más me gustó fue la piscina y nadaba todos los días.

El mejor día fue cuando dimos una vuelta en bici por las montañas. Fue muy duro pero merecía la pena porque las vistas eran increíbles. Llevamos un picnic para mediodía y comimos al lado de un lago rodeado de montañas.

El año próximo, me gustaría ir a Australia con mi familia. Tenemos amigos allí y queremos visitarles. No he estado nunca y me parece que sería una buena experiencia conocer partes diferentes del país.

Last year I went on holiday to Italy with my family and my best friend. We travelled by car and the journey lasted three days altogether. We went through the Channel Tunnel and then we travelled through France to the Alps. We arrived in Italy very tired, but very happy.

We stayed on a lovely campsite near an enormous lake. There were good facilities and the countryside was beautiful. What I liked the most was the swimming pool and I swam every day.

The best day was when we went for a bike ride in the mountains. It was very hard, but worth it because the views were incredible. We took a picnic for lunch and we ate next to a lake surrounded by mountains.

Next year, I would like to go to Australia with my family. We have friends there and we want to visit them. I have never been and I think that it would be a good experience to get to know different parts of the country.

Free Time and Leisure

Pages 40–41 Leisure and Pastimes
Test Yourself Answers
1 Soaps are my favourite programmes.
2 In the evening / At night I prefer to surf the Internet.
3 Me encanta ir al cine con mis amigos.
4 La semana pasada fui al teatro con mi primo.

Stretch Yourself Answers
1 Los documentales son más interesantes que las noticias.
2 Las emisiones deportivas son menos aburridas que los dibujos animados.

Pages 42–43 Sports and Exercise
Test Yourself Answers
1 I go skateboarding in the park.
2 I play table tennis every day at school.
3 Suelo practicar el boxeo el fin de semana.
4 El sábado voy a ir de pesca con mi hermano.

Stretch Yourself Answers
1 Practico el patinaje sobre hielo desde hace dos años y tengo que ir a la pista de hielo dos veces a la semana.
2 Practicaba el atletismo desde hacía cinco años.

Pages 44–45 More Free-time Activities
Test Yourself Answers
1 I don't know how to play an instrument but I love listening to music.
2 I always read fashion magazines because I'm interested in clothes.
3 Veo la tele para relajarme.
4 Para mí es muy aburrido.

Stretch Yourself Answers
1 El sábado por la mañana íbamos a la piscina y yo nadaba con mis hermanos.
2 Cuando era joven, tocaba la trompeta.

Pages 46–47 Clothes and Fashion
Test Yourself Answers
1 Have you got the skirt in a size 38?
2 Yesterday I bought a gold necklace.
3 un reloj de oro una bufanda de lana unas medias negras
4 Ayer compró un vestido verde y unas botas marrones.

Stretch Yourself Answers
1 ¿Te gustan mis zapatos nuevos? ¡Los compré porque fueron baratos!
2 ¿Te gusta este cinturón de cuero? Lo compré para mi hermano.

Page 48–49 Shops and Services
Test Yourself Answers
1 Where is the checkout, please?
2 Yesterday we sent a parcel to my grandmother.
3 Mi padre es carnicero, mi madre es panadera.
4 Voy a rellenar un formulario en correos.

Stretch Yourself Answers
1 Me gusta esta postal pero ésa no.
2 ¿Cuál pastelería? ¿Esta o aquella?
3 ¿Quieres ir a esta tienda de recuerdos o a ésa?

Page 50–51 Food and Drink
Test Yourself Answers
1 Give me 150g of ham.
2 For me, an orange juice, please.
3 Deme un kilo de plátanos y medio kilo de judías verdes, por favor.
4 Mi amigo siempre come demasiado.

Stretch Yourself Answers
1 El camarero habla más rapidamente que yo.
2 En el bar, pido un café con leche más a menudo que un refresco.

Pages 52–53 The Perfect and Pluperfect Tenses
Test Yourself Answers
1 This morning I've had cereal with milk for breakfast. (lit. I have breakfasted)
2 He / She had asked me earlier if I wanted to go to the cinema.
3 Han escrito una carta a mi profesor.
4 Ya había visitado Sevilla.

Stretch Yourself Answers
1 ¡Al final, el pasaporte que había perdido estaba en su bolsa!
2 Acababa de comer cuando llegó a casa su padre.

Pages 54–55 Eating Out
Test Yourself Answers
1 For the first course, I'll have Spanish omelette, please.
2 I usually have toast with strawberry jam for breakfast.
3 Para mí, de primero el chorizo y de segundo el cocido, por favor.
4 ¿Te apetecen los mariscos o la carne?

Stretch Yourself Answers
1 Me encanta la paella de mariscos pero ¡no me gustan nada las albóndigas!
2 Me gusta mucho el bistec pero ¡no me gusta nada el pollo!

Pages 56–57 Healthy Living
Test Yourself Answers
1 I am totally against drugs. They're very dangerous!
2 Smoking is very harmful to your health.
3 Para llevar una vida más sana, no deberías fumar.
4 Hay que comer bien y hacer ejercicio para estar en forma.

Stretch Yourself Answers
1 Está claro que tomar drogas es peligroso.
2 A mi modo de ver, hoy en día, las drogas son un problema muy grave.

Page 58–59 Illness and Accidents
Test Yourself Answers
1 I feel ill – my throat hurts a lot.
2 Yesterday I burnt my finger and it hurt quite a lot.
3 Me duelen las piernas y los pies.
4 Tengo que guardar cama porque tengo fiebre.

Stretch Yourself Answers
1 Estoy guardando cama porque no me siento bien.
2 Ayer estaba cocinando en la cocina cuando me desmayé.

Page 60–61 Celebrations and Special Events
Test Yourself Answers
1 They gave me a leather jacket as a present.
2 My mum made a delicious dinner / evening meal.
3 Celebro mi cumpleaños con mis amigos.
4 Espero con ilusión el Año Nuevo Chino.

Stretch Yourself Answers
1 A menudo celebramos la Pascua en casa de mis tíos.
2 Voy a celebrar mi cumpleaños dentro de poco.

Pages 62–63 Answers to Practice Questions
(See the guidance on page 90.)
1 **a)** Plays basketball
 b) Ana
 c) Goes to football matches
 d) Ana
 e) Athletics training
 f) Something to eat / drink
 g) To a lake (near the town)
 h) It depends if he has enough money.
2 **A** e)
 B c)
 C b)
3 Example answers:
 a) Suelo ir de compras a los centros comerciales porque creo que hay más variedad y también, ¡no tienes que salir a la calle cuando hace frío! Antes siempre iba a las tiendas de diseño, pero ahora me parecen demasiado caras.
 I usually go shopping in shopping centres because I think that there is more variety and also, you don't have to go out into the street when it's cold! Before, I always used to go to designer shops, but now they seem too expensive.
 b) Depende. Normalmente prefiero ir de compras con mis amigos porque siempre lo pasamos bien. Pasamos todo el día allí, vamos de tienda en tienda, comemos juntos y nos reímos mucho. Es un día muy divertido.
 It depends. I usually prefer to go shopping with my friends because we always have a good time. We spend the whole day there, we go from shop to shop, we eat together and we laugh a lot. It's a really fun day.
 d) La última vez que fui de compras, fue con mi madre y mi hermana. Buscábamos un traje para una boda y pasamos un día entero en el mismo almacén. Comimos allí a mediodía y lo pasamos muy bien. Al final las tres compramos un traje nuevo y nos quedamos muy contentas.
 The last time I went shopping, it was with my mum and my sister.

We were looking for an outfit for a wedding and we spent a whole day in the same department store. We ate there at lunchtime and we had a good time. In the end, we all bought a new outfit and we were very happy.

e) En mi opinión, los grandes almacenes son buenos porque con todos los departamentos diferentes, es como si hubiera muchas tiendas juntas. Sin embargo, a veces pueden ser más caras, depende de donde vas.
In my opinion, big department stores are good because, with all of the different departments, it's as if there were lots of shops together. However, sometimes they can be expensive, it depends where you go.

4. Example answer:

Suelo comer una dieta equilibrada, o sea, verduras, fruta, carne, pescado, etc. Me gustan mucho los dulces y tengo que intentar no comer demasiados pero creo que más o menos como bien. Me gusta beber agua y refrescos y a veces tomo té por la mañana.

Cuando era pequeño hacía más deporte que ahora porque después del colegio estaba siempre jugando al fútbol o al cricket en el jardín con mis hermanos. Ahora, no tengo tiempo porque tengo muchos deberes todas las noches. También comía muchos dulces, pero ahora no los como tanto.

En el futuro tendré que continuar con el deporte y con una dieta equilibrada. Lo que sí necesito hacer es dormir por lo menos ocho horas porque a veces no me acuesto hasta muy tarde y me siento muy cansado el día siguiente.

En mi opinión tomar drogas y fumar son los dos asquerosos. No aguanto el humo del tabaco y no me gusta estar en lugares cuando fuman los demás. Tomar drogas me da mucho miedo y no lo haré nunca. En cuanto al alcohol, creo que si solo bebes un poco, no hay problema.

I usually eat a balanced diet, that is, vegetables, fruit, meat, fish, etc. I really like sweet things and I have to try not to eat too many but I think that I eat well more or less. I like to drink water and soft drinks and I sometimes have tea in the morning.

When I was little I used to do more sport than now because after school I was always playing football or cricket in the garden with my brothers. Now, I don't have time because I have a lot of homework every night. Also I used to eat a lot of sweet things but now I don't eat them as much.

In the future, I will have to continue with the sport and with a balanced diet. What I do need to do is to sleep at least eight hours because sometimes I don't go to bed until very late and I feel very tired the next day.

In my opinion, taking drugs and smoking are both disgusting. I cannot stand cigarette smoke and I do not like being in places when others are smoking. Taking drugs scares me and I will never do it. As for alcohol, I think that if you only drink a little, there isn't a problem.

Education and Work

Page 64–65 School and School Subjects
Test Yourself Answers

1 I don't like physics because I don't understand anything.
2 I'm going to fail music because I always get bad marks.
3 Quiero sacar buenas notas en inglés.
4 Me encantaría estudiar el italiano.

Stretch Yourself Answers

1 Odio las ciencias porque son complicadas pero me interesan las matemáticas.
2 A mi amigo y a mí nos encanta el teatro porque es divertido.

Page 66–67 Talking about School
Test Yourself Answers

1 Lessons start at quarter to nine.
2 In my primary school we used to wear a uniform but now it's not compulsory.
3 Odio mi uniforme escolar porque es muy feo.
4 No me gusta la química porque siempre tenemos muchas pruebas.

Stretch Yourself Answers

1 En mi colegio siempre se debe escuchar al profesor y respetar a los demás.
2 No se permite llevar maquillaje ni joyas.

Pages 68–69 Pressures and Problems at School.
Test Yourself Answers

1 Bullying happens quite a lot in my school.
2 I am fed up with so many exams.
3 Hay que estudiar mucho para aprobar los exámenes.
4 Creo que voy a suspender el examen de inglés.

Stretch Yourself Answers

1 El profesor habló con el estudiante en cuanto a su comportamiento.
2 El acoso escolar es, sin duda, uno de los mayores problemas en los institutos.

Pages 70–71 Part Time jobs and work Experience
Test Yourself Answers

1 I did my work experience in a shop.
2 I think it was a good experience.
3 Hago de canguro los sábados por la noche.
4 En mi opinión la experiencia fue muy útil.

Stretch Yourself Answer

1 El año pasado hice mis prácticas en un banco y fue muy interesante. Trabajé en la caja, sirviendo a los clientes, y me gustaría trabajar allí en el futuro.

Pages 72–73 Future Career Plans
Test Yourself Answers

1 I want a fun job.
2 I want to be a journalist.
3 Mi amigo quiere un trabajo útil.
4 Tengo la intención de viajar con mi hermana.

Stretch Yourself Answers

1 Si apruebo mis exámenes, continuaré estudiando.
2 Si no voy a la universidad, trabajaré con mi padre.

Pages 74–75 Answers to Practice Questions
(See the guidance on page 90.)

1 **a)** Elena P+N
b) Paco P
c) Miguel P+N

2 b), d), e), h)

3 Example answers:

a) Mis asignaturas preferidas son los idiomas porque para mí son muy interesantes y creo que es muy importante intentar comunicarse en el idioma del país cuando voy de vacaciones. También me interesa mucho el deporte así que me encantan mis clases de educación física.
My favourite subjects are languages because, for me, they are really interesting and I think that it's very important to try to communicate in the language of the country when I go on holiday. I am also very interested in sports so I love my PE lessons.

d) Diría que lo más estresante en el instituto es lo de siempre intentar sacar buenas notas. Tengo que hacer muchos deberes todos los días y casi nunca tengo tiempo libre. Cuando era más pequeño nunca tenía esta cantidad de trabajo; a veces es demasiado.
I would say that the most stressful thing at school is always trying to get good marks. I have to do lots of homework every day and I almost never have free time. When I was younger I never had this much work; sometimes, it's too much.

e) A mí no me gusta mi uniforme escolar porque no es muy bonito. Tenemos que llevar una corbata, que no me gusta nada y el color es muy oscuro. Si pudiera escoger, yo llevaría mi propia ropa.
I don't like my school uniform because it's not very nice. We have to wear a tie that I don't like at all and the colour is very dark. If I could choose, I would wear my own clothes.

f) Yo en general, estoy muy contenta con mi vida escolar. Tengo muchos amigos en el instituto y la mayoría de mis clases están bien y los profes también. Me quejo a veces de los deberes pero realmente no hay ningún problema.
In general I am very happy in my school life. I have a lot of friends at school and the majority of my classes are good and the teachers as well. I complain at times about the homework but really there isn't a problem.

4 Example answer:

El año pasado hice mis prácticas laborales en un banco en la ciudad y me gustó mucho. Lo hice allí porque en el futuro creo que quiero ser contable ya que me encantan las matemáticas.

Tenía que trabajar con los empleados en la oficina, a veces con el ordenador, otras veces ordenando papeles y ayundándoles con tareas diferentes. Un día trabajé un poco en la caja pero realmente no podía hacer mucho pues no tengo experiencia.

Yo diría que fue una buena experiencia y que lo pasé bien. Pienso que aprendí bastante, y me ha ayudado a decidir que eso es lo que quiero hacer en el futuro.

En el futuro, voy a ir al colegio para continuar con mis estudios. Después, espero ir a la universidad para estudiar matemáticas y mientras estudio, quiero buscar un trabajo a tiempo parcial, preferiblemente en un banco, para ganar un poco de dinero y experiencia también.

Last year I did my work experience in a bank in the city and I really liked it. I did it there because in the future I think that I would like to be an accountant as I love maths.

I had to work with the employees in the office, sometimes on the computer, other times organising papers and helping with different tasks. One day, I worked a little on the cash desk but I could not really do a lot as I do not have the experience.

I would say that it was a good experience and that I enjoyed it. I think that I learnt quite a lot and it has helped me to decide that that is what I want to do in the future.

In the future, I am going to go to college to continue my studies. Afterwards, I hope to go to university to study maths and while I am studying, I want to look for a part-time job, preferably in a bank, so that I can earn a bit of money and also some experience.

The Wider World

Pages 76–77 The Subjunctive and the Conditional
Test Yourself Answers

1 It's important that you go with me.
2 Would you be able to go out tonight?
3 ¡Quiero que hables español!
4 Bebería té.

Stretch Yourself Answers

1 Es una pena que no podamos hacer más.
2 ¿Podrías reciclar basura y apagar las luces, por favor?

Pages 78–79 Environmental Issues
Test Yourself Answers

1 My city is very polluted.
2 One / You should get showered / take a shower instead of having a bath.
3 Deberíamos consumir menos energía.
4 Es mejor ir andando / a pie que ir en coche.

Stretch Yourself Answers

1 Las ciudades están estropeadas por tanto tráfico.
2 El planeta será destruido por la contaminación.

Pages 80–81 Global Issues
Test Yourself Answers

1 We have to search for renewable energy.
2 For me, drought is very worrying.
3 Lo que más me preocupa es el efecto invernadero.
4 Tenemos que aprender a crear un mundo mejor.

Stretch Yourself Answers

1 La persona que más me preocupa es mi hermana.
2 Lo que tenemos que hacer es mejorar la situación.

Page 82–83 Social Issues
Test Yourself Answers

1 I want to become a volunteer.
2 I am trying to contribute more.
3 Hay una falta de respeto en la sociedad.
4 Hay que educar a la gente.

Stretch Yourself Answers

1 Es nuestra responsabilidad de mejorar la sociedad.
2 Todo el mundo tiene sus derechos humanos.

Page 84–85 Technology
Test Yourself Answers

1 Emails are quicker.
2 I think that chatting on the Internet is very dangerous.
3 Navego por Internet todos los días.
4 Uso mi móvil para mandar mensajes.

Stretch Yourself Answers

1 ¿Es ése tu portátil? No, es suyo.
2 ¿Los móviles? El suyo está muy bien. Mejor que los nuestros.

Pages 86–87 Answers to Practice Questions
(See the guidance on page 90.)

1 **a)** Five from:
They don't recycle.
They don't reuse anything.
They leave lights on.
They don't separate the rubbish.
Mother has a bath every night.
Father takes her to school in car.
b) Two from:
Explain the problem.
Tell them how important it is to her.
Tell them she wants them to help her.
c) Put rubbish into correct containers.
Reuse bags when shopping.

2 **a)** C **b)** D **c)** E **d)** A **e)** F **f)** B

3 Example answers:

a) El tema del medio ambiente me preocupa mucho. Es un problema enorme y creo que no estamos haciendo lo suficiente para mejorar la situación. Desafortunadamente hay gente a quien no le importa el calentamiento global y es difícil convencerla de que hay que actuar.
The topic of the environment worries me a lot. It's an enormous problem and I think that we're not doing enough to improve the situation. Unfortunately there are people who don't care about global warming and it's difficult to convince them that they have to act.

c) En el futuro, me gustaría ver más centros de reciclaje en todas las ciudades. Los gobiernos locales deben invertir más dinero en otros métodos de producir energía como la energía eólica. Es imprescindible que todos hagamos lo más posible para mejorar la situación.
In the future, I would like to see more recycling centres in all cities. Local governments should invest more money in other ways of producing energy, such as wind power. It's essential that we all do as much as possible to improve the situation.

d) En mi pueblo el tráfico causa muchos problemas como la contaminación del aire, que es muy serio. En el pasado no había mucho tráfico pero ahora todo el mundo viaja en coche y muchas familias tienen dos coches cada una. Hay que usar más el transporte público.
In my town, traffic causes a lot of problems such as air pollution, which is very serious. In the past, there wasn't a lot of traffic but now everyone travels by car and lots of families have two cars each. We must use public transport more.

f) Los problemas del medio ambiente afectan al mundo entero.Tenemos que trabajar juntos para buscar alternativas a los combustibles fósiles. No es suficiente decir que un problema en otro país no nos afecta aquí en nuestro país, como por ejemplo, la deforestación. ¡Hay que salvar el planeta!
Environmental problems affect the whole world. We have to work together to look for alternatives to fossil fuels. It's not sufficient to say that a problem in another country doesn't affect us here in our country, such as, for example, deforestation. We must save the planet!

4 Example answer:

Uso la informática cada vez más en el colegio en mis clases y para hacer los deberes, etc. A menudo hacemos los deberes en casa con el ordenador y los mandamos a nuestros profesores por correo electrónico. También me gusta chatear con amigos y descargar música del Internet.

En el pasado, mis padres nunca usaban los ordenadores para hacer las cosas que hago yo. En el cole escribían todo en los cuadernos y nadie tenía ordenadores en casa. Ha cambiado muchísimo en los últimos veinte años.

Para mí, lo negativo de usar la informática para hacer todo es que muchos jóvenes no salen nunca de su casa para quedarse con sus amigos porque comunican por el ordenador. Yo creo que esto va a causar problemas sociales en el futuro.

En el futuro, pienso que los ordenadores van a hacer muchos trabajos que en el pasado hacíamos nosotros. Va a haber menos empleos para la gente a causa de lo que podrán hacer los ordenadores.

I use ICT more and more often at school in my lessons and for doing homework. Often we do homework at home on the computer and then send it to our teachers by email. I also like to chat to my friend and download music from the Internet.

In the past, my parents never used computers to do the things that I do. At school they used to write everything in their exercise books and nobody had computers at home. It has changed so much in the last 20 years.

For me, the negative side of using ICT to do everything is that lots of young people never leave their house to meet up with their friends, because they communicate through the computer. I think that this is going to cause social problems in the future.

In the future, I think that computers are going to do more jobs that, in the past, we used to do. There are going to be less jobs for people because of what computers will be able to do.

Index